T0201875

Science, Secrecy, and the Smithsonian

Science, Secrecy, and the Smithsonian

The Strange History of the Pacific Ocean Biological Survey Program

ED REGIS

OXFORD
UNIVERSITY PRESS

OXFORD
UNIVERSITY PRESS

Oxford University Press is a department of the University of Oxford. It furthers
the University's objective of excellence in research, scholarship, and education
by publishing worldwide. Oxford is a registered trade mark of Oxford University
Press in the UK and certain other countries.

Published in the United States of America by Oxford University Press
198 Madison Avenue, New York, NY 10016, United States of America.

© Oxford University Press 2022

All rights reserved. No part of this publication may be reproduced, stored in
a retrieval system, or transmitted, in any form or by any means, without the
prior permission in writing of Oxford University Press, or as expressly permitted
by law, by license, or under terms agreed with the appropriate reproduction
rights organization. Inquiries concerning reproduction outside the scope of the
above should be sent to the Rights Department, Oxford University Press, at the
address above.

You must not circulate this work in any other form
and you must impose this same condition on any acquirer.

Library of Congress Cataloging-in-Publication Data
Names: Regis, Edward, 1944- author.
Title: Science, secrecy, and the Smithsonian : the strange history of the
Pacific Ocean Biological Survey Program / Ed Regis.
Other titles: Strange history of the Pacific Ocean Biological Survey Program
Description: New York, NY : Oxford University Press, [2022] |
Includes bibliographical references.
Identifiers: LCCN 2022018862 (print) | LCCN 2022018863 (ebook) |
ISBN 9780197520338 (hardback) | ISBN 9780197520352 (epub) |
ISBN 9780197520369
Subjects: LCSH: Biological weapons—United States—Testing—History—
20th century. | Biological weapons—Environmental aspects—Pacific Area. |
Project 112 (U.S.)—History. | Birds—War use—History—20th century. |
Birds—Immunology. | Pacific Ocean Biological Survey Program—History. |
Offenses against the environment—Government policy—United States.
Classification: LCC UG447.8 .R447 2022 (print) | LCC UG447.8 (ebook) |
DDC 358/.3882—dc23/eng/20220603
LC record available at https://lccn.loc.gov/2022018862
LC ebook record available at https://lccn.loc.gov/2022018863

DOI: 10.1093/oso/9780197520338.001.0001

1 3 5 7 9 8 6 4 2

Printed by Sheridan Books, Inc., United States of America

To Pamela Regis

Contents

Preface

Between 1962 and 1969, the Smithsonian Institution, whose foundational mission was to promote "the increase and diffusion of knowledge among men," undertook a large-scale biological survey of a group of uninhabited tropical islands in the Pacific. This would turn out to be one of the largest and most sweeping biological survey programs ever conducted. It was a seven-year-long enterprise during the course of which Smithsonian personnel banded some 1.8 million seabirds; captured hundreds of live specimens; took blood, liver, and spleen samples from hundreds of others; and cataloged the avian, mammalian, reptile, insect, and plant life of forty-eight Pacific islands, atolls, and pinnacles, some of which had never before been explored by scientists.

But there was a twist. The survey was initiated not by the Smithsonian Institution itself but by officials of the Deseret Test Center in Utah, a recently established US Army command post responsible for planning, testing, and evaluating biological and chemical weapons systems. Further, the project was funded and would be overseen from the start by the US Biological Laboratories at Fort Detrick, Maryland, home of the US biological warfare program. In signing the contract to perform the survey, the Smithsonian Institution became a literal subcontractor to a top-secret biological weapons project.

This meant that for the lifetime of the program, the Smithsonian contractually subordinated a portion of its time, attention, and activities to the wants, needs, and dictates of the US Army. More important, by participating in the survey, the Smithsonian scientists helped prepare the way for a number of the army's secret biological weapons trials in the Pacific. In fact, while the survey was still in progress, a team of US biological weapons developers traveled to a few of the islands and open ocean areas the institution's scientists had explored and documented, and released into the open air two different pathogenic, incapacitating biological warfare agents and one lethal agent.

Critics later charged the Smithsonian with having entered into a Faustian bargain that made the institution complicit in the sordid business of germ warfare, a form of combat that, if it were ever put into practice against human populations, could cause disease, suffering, and death on a mass scale. The Smithsonian had no proper role supporting any such activities, the critics said, and should never have agreed to perform an army-financed biological survey to begin with.

Still, it was clear that the Smithsonian had a lot to gain by participating in the study, for this was an unparalleled opportunity for its scientists to gather firsthand knowledge of an area of the world whose birdlife was until then only poorly known and understood. And it would be doing so on a grand scale: the Pacific Ocean is immense, its dimensions colossal. At its widest it stretches ten thousand miles from east to west and another ten thousand from north to south. In marked contrast to this immensity, however, is the minute size of the islands scattered in a seemingly random pattern across its vast surface. Typically, these islands, atolls, rocky outcroppings, and other bits of land are no more than a few hundred acres in area. One of them, Kingman Reef in the North Pacific, approximately halfway between Hawaii and American Samoa, is barely seven acres in size, "smaller than most ships," according to one observer.

The prospect of Smithsonian scientists visiting such tiny, remote, and barren tropical islands, on multiple occasions over a period of years, with all travel, salaries, and other expenses paid by an outside agency—this was an opportunity that the Smithsonian's administrators apparently could not pass up. After all, it was the chance for its scientists to learn in detail what life forms inhabited these isolated locations and to publish their findings freely and openly, with virtually no strings attached.

However, all of this was in the face of the further fact that, at least at first, the army did not make clear to Smithsonian administrations why its biological weapons developers wanted the knowledge that their scientists were being asked to provide. Had they been told the full truth about the army's motivation at the outset, the institution's upper echelons might have seen the project in an entirely different light.

Later, after news of the survey program became public, critics advanced a cynical, indeed sinister, explanation of the army's interest in collecting information about Pacific seabird distribution and migration patterns. They claimed that the army was contemplating the use of birds as vectors or carriers of diseases against potential enemy populations. It was common knowledge that birds can harbor viruses, bacteria, or other agents, on their feathers, in their bloodstreams, or in their digestive tracts. In theory, then, it would be possible to deliberately infect birds with a disease-causing biological agent.

And if it was also known that a given bird species tended to migrate to specific population centers, then it would be possible to use them to spread diseases among the area's residents. As one journalist put it, the army wanted to use "one of God's gentlest creatures, a gull . . . for a doomsday assignment," converting birds into bioweapons delivery systems, into "bird bombs."

It was for this reason, the critics alleged, that the army wanted the Smithsonian scientists to band so many birds: the recovery of banded birds was the only sure way of knowing where members of a given species tended to migrate. With that knowledge, birds of that species could be intentionally infected with pathogenic microbes that the birds could carry to their natural destinations, thereby infecting humans.

This was *not* in fact the army's true motivation for initiating and sponsoring the Smithsonian's Pacific Project, but the institution's administrators did not know that at the time, nor is it clear that they ever even asked. The Smithsonian was in the business of generating and disseminating knowledge. Acquiring that knowledge took money, and here was the army offering it.

"To me, as a bird man, it was a wonderful breakthrough because it was a source of funds," said S. Dillon Ripley, the Smithsonian's secretary during the bulk of the project. "That's all I know about it." Or all he cared to know, apparently.

* * *

Whatever the plausibility of the "bird bomb" or bioweapons delivery system hypothesis, it is undeniable that the army's participation in the Pacific island survey project led to some strange, and even somewhat bizarre, encounters between Smithsonian scientists and US Army biological weapons experts. For it was not only *knowledge* that the army wanted from the Smithsonian's fieldworkers—far from it. The army's biological warfare laboratories were staffed by scientists, and these scientists wanted much more than just reams of naked bird migration, distribution, and population data. They also wanted *specimens*, first in the form of live birds that were to be captured on certain islands, shipped back to the US mainland, and then transported to army biological warfare research laboratories at Fort Detrick and Dugway Proving Ground, southwest of Salt Lake City.

The Smithsonian's field team members were also asked to provide other specimens to bioweapons scientists, who would analyze their stomach contents and discover what the birds fed upon. They were asked to collect bird brains and eggs, as well as ectoparasites (such as fleas or ticks) living on avian skins and feathers. They were even asked to collect sea water and ocean plankton samples, although for what precise purpose was unclear.

The army's need for physical samples and specimens led to a type of professional collaboration that one does not normally associate with the disinterested and dispassionate search for pure knowledge that typically motivated Smithsonian researchers. In 1963, early on in the program, Binion Amerson, one of the Pacific Project's most valuable, talented, and productive scientists, was stationed on Sand Island in Johnston Atoll, about seven hundred miles southwest of Honolulu. While there he was joined by several high-ranking members of the army's biological warfare project. One was from Fort Detrick; the others had come from the army's Deseret Test Center (DTC).

In his typed field notebook entry for July 24, 1963, Binion Amerson recorded:

Northwest MATS Flight brought Mr. Bill Miller, Virologist, Fort Detrick, Maryland, for a week's stay. He brought with him some of the blood taking equipment (syringes, etc.) not included in the trunk. . . . Also on the plane were Mr. John Bushman, Col. Smith and Mr. Henderson from the Deseret Test Center. . . . I discussed various points with Mr. Bushman. Among them was security. He assured me they were working on it. . . .

We discussed with Mr. Miller the upcoming bird and blood sample shipment. Two hundred fifty Wedgetailed Shearwater are wanted during the middle or last of September. No other birds will be needed by Detrick until March. Approximately one hundred blood sera samples are wanted on August 13. Mr. Bushman will arrange for dry ice to be shipped here on August 12.

Later, in his entry for August 14, 1963, Amerson wrote:

Chief Gragosian from DTC arrived today on the N. W. MATS flight. He brought some supplies with him for collecting some of the items he is to take back. These items are, in order of precedence:

200 Live Shearwaters
Miller's Sera
100 Bird Sera (Shearwater)
100 Bird Spleen (Shearwater)
Ectoparasites (Ticks)

100 Mouse Sera
100 Mouse Spleen

By any standard, this was a strange business. Here was a Smithsonian Institution scientist collaborating with a biological weapons virologist and several Deseret Test Center project officers, about matters of "security" (for this was a secret project for the army) and about the taking of live birds, blood samples, and other biological specimens from Johnston Atoll, and shipping them in boxes to US Army germ warfare laboratories on the mainland.

Such cooperative associations between Smithsonian scientists and army biological warriors were by no means routine, but they were not uncommon, either, and occurred on multiple occasions. The fact that they occurred at all was remarkable, as was the fact that the Smithsonian field teams traveled from island to island aboard US Navy ships, in many cases on the very same light tugboats as well as on larger transport craft that the army used in its biological weapons trials at sea.

And in at least one case, a Smithsonian scientist actually participated in one of those biological weapons trials.

The US Army's role in the Pacific Project is an inseparable part of the story: the army originated the project, financed it, and used the knowledge generated by the Smithsonian's scientists. This conflux of circumstances raised some obvious questions. What exactly did the army want with so many live birds—so many in some cases that the scientists complained to Deseret Test Center officers that some of these takings were significantly depleting a given island's bird populations? And why this preoccupation with shearwaters specifically as opposed to birds of other species? What use did the army have for mouse sera and spleens and with the massive amounts of other samples and specimens that the Smithsonian fieldworkers collected over the project's seven years?

Any attempt to answer these and other questions satisfactorily, or to gain full information about what the Smithsonian's Pacific Project was really all about, is circumscribed by two factors operating at cross-purposes. One is the open nature of the Smithsonian's records, field notes, data books, internal reports, and publications relating to the project. All these documents and more are freely available to the public, although many are accessible only at the Smithsonian Institution Archives. As such, they provide a virtually complete historical record of what the project's scientists did and what they discovered on their numerous forays into the Pacific. What is missing from these accounts, however, is exactly *why* they were doing what they did, what the precise rationale was for all of their painstaking data gathering, the capturing of live birds, and the biological sample collecting.

Working against the Smithsonian's openness is the classified and secret status of the Army's records, documents, and objectives relating to its biological warfare projects, research, and field trials.

Of necessity, any historian of the United States' (or other nation's) germ warfare program is a trafficker in redacted documents, with paragraph after paragraph and page after page heavily blotted out in black ink or "intentionally left blank." This means that the army's motives in fostering the Pacific Project, and what it wanted with any given item of information, are often matters of inference from surviving scraps of paper containing, in some cases, a few lines of barely readable text. Parts of the story therefore remain obscure, comprising what might be called the program's dark or invisible history.

But for all the military's attempts at secrecy, obfuscation, and denial, a record of past events is nevertheless preserved in the form of transcripts of older personal interviews with project participants; their personal papers, photographs, published and unpublished memoirs; and even in the rare cases of mistakenly unredacted, but quite telling, paragraph or page. From such sources, and from what already exists in the open scientific literature that arose from the project, it is possible to piece together and reconstruct a substantive, connected, and reliable account of the Pacific Project, its visible history.

The project was formally and finally known as the Smithsonian Institution's Pacific Ocean Biological Survey Program—POBSP for short. It is one of the strangest and, even today, one of the most mysterious endeavors ever mounted by the Smithsonian. It is the unraveling of its mysteries, and the story of that project and its consequences, that will be told in this book.

1

Secrecy Comes to the Smithsonian

In the fall of 1962, at the height of the Cold War, officers representing the three main military services, the US Army, Navy, and Air Force, arrived at the Smithsonian Institution in Washington. The officers were from the Deseret Test Center (DTC), a new military installation in Fort Douglas, Utah.

Fort Douglas, on the eastern edge of Salt Lake City, was established during the Civil War as a small military installation. It grew through the years and served a variety of purposes during the United States' series of foreign wars. By World War I, a collection of buildings had arisen at the site, and they were used, among other purposes, as an internment camp for German citizens living in the United States and for German prisoners of war.

Never very large in terms of land area, the new Deseret Test Center was staffed at its peak by 227 military and civilian personnel, a mouse as compared to the elephant in the room: Dugway Proving Ground, some sixty miles to the southwest. Dugway was an immense presence in Utah.

Located at the southern end of the Great Salt Lake Desert, Dugway Proving Ground came into existence in the aftermath of the Pearl Harbor attack, when the United States decided to drastically augment its military capabilities. It needed more and better weapons, and a greater variety of them. The proving ground was created on February 6, 1942, at the order of President Franklin D. Roosevelt, who withdrew 126,720 acres of land (about two hundred square miles) from the public domain and gave it over for use by the War Department. The place was named after the Dugway Mountains to the east, which were themselves named for the result of digging a trench, or "dugway," along a hillside to keep covered wagons from tipping over. What the army was "proving" at the ground were weapons—not only conventional explosive munitions, but also chemical, toxin, and biological weapons and their associated delivery systems.

Dugway scientists often disseminated their biological and chemical agents in open-air field trials. To contain these hazardous substances, and to prevent them from affecting people living in adjacent areas, the proving ground had to be big, and so the army kept expanding its size, increasing its area first to

634 square miles and finally to more than 1,200 square miles, an expanse of land roughly the size of Rhode Island.

Between World War II and the time that the Deseret Test Center officers paid a visit to the Smithsonian, Dugway's weapons developers had tested a large variety of both incapacitating and lethal chemical and biological agents and devices. Among the chemicals the researchers experimented with was the nerve agent VX. VX was so incredibly toxic that less than ten milligrams—a droplet about the size of a dime—on the skin was said to be sufficient to kill an adult in fifteen minutes.

Scientists at Dugway also experimented with a variety of biological agents, including, in 1955, an aerosol cloud of the causative agent of Q fever, *Coxiella burnetii*, which was dispersed over a group of thirty human volunteer test subjects. They were Seventh-day Adventists who, because of their religious beliefs, remained noncombatants but had no objection to being used as guinea pigs.

Virtually everything that went on at Dugway was done under wraps, and the reports on the tests and their results were marked Secret at the time, and for the most part have remained so.

In 1962, when the DTC officers made their trip to the Smithsonian, both Deseret and Dugway were in the business of chemical and biological warfare preparations in earnest. Despite its name, however, no actual testing work was done at the Deseret Test Center. Its personnel were engaged in the planning and evaluation of such tests, not in their execution. The final objective of the facility was expressed symbolically in the DTC logo, which featured an image of the earth partially obscured by aerosol clouds of biological or chemical agents wafting over the surface of the planet. It was not a pretty picture.

* * *

The Deseret Test Center's representatives met with Remington Kellogg, the Smithsonian Institution's assistant secretary, and director of the United States National Museum, on the Mall in Washington. Kellogg was a natural historian in the classic sense and was interested in a wide variety of birds and mammals. He held undergraduate degrees from the University of Kansas, had done graduate work in zoology at the University of California, Berkeley, and wrote his PhD thesis on the history of whales. He was a lifelong whale connoisseur, and at the Smithsonian he presided over the installation of a ninety-four-foot-long model of a blue whale in the museum's Hall of Life in the Seas. Nick Pyenson, a curator at the museum, called Kellogg "arguably the dean of whale biology in the United States."

The officers from Fort Douglas explained to Kellogg that they were interested in having a massive survey undertaken of the populations and migration patterns of the birds and mammals across a large area of the tropical Pacific. The air force officer said that they wanted this information to help prevent bird strikes with aircraft on the Pacific islands from which they operated. This was a plausible explanation: aircraft strikes with birds at the US Naval Air Station Midway Island, for example, had been a problem of major dimensions. The navy, for its part, said that it would participate in the survey effort by transporting field crew members from island to island.

But it was the army that had the most important and pressing reason for making the survey, although exactly how much the representative told to Remington Kellogg, and when, if ever, is unclear. Essentially, the army wanted to find out which of the islands would be most suitable for a series of large-scale biological warfare trials it was planning in the Pacific. The new tests being contemplated were to be performed over so wide a geographical area that even Dugway Proving Ground, big as it had become, was insufficient to safely contain the pathogens that would be used.

In the process of disseminating the pathogenic agent clouds, however, the army wanted to minimize any potentially negative effects on the ecology of the islands and also wanted to avoid introducing diseases not already endemic to the birds and mammals of the regions in question. Those were operational constraints upon the army's fulfillment of its ultimate goal, which was to determine how effective certain bacteriological agents would be when spread by airborne means over a wide area of open ocean, or even, perhaps, when carried by certain animal species as vectors of disease. To remain within the limits of those constraints, the army needed to know the migration patterns of the birds and mammals in the areas of the Pacific where its biological weapons tests would be conducted.

There were a few restrictions that the Deseret officers attached to the project they described to Kellogg, one of which was that the biological weapons testing details be kept secret by whatever organization did the survey. Another was that all of the personnel taking part in the operation would have to obtain Secret security clearances, as well as to be immunized against certain unspecified diseases. A third was that although the information generated by the survey could be freely published in the open scientific literature, the contents of any such publications would first have to be cleared by the relevant army authorities.

And so the officers asked Kellogg to recommend one or more universities whose ornithological faculty members might be up to the rather imposing task they had outlined. Instead, Kellogg suggested that the Smithsonian itself was ideally qualified to take on the project. The institution had a long record of experience in the scientific study of birds, it possessed a large ornithological library, and it already housed one of the world's largest collections of bird skins. It had on hand a staff of competent ornithologists, in addition to which any number of others could be hired on as needed. Further, the Smithsonian was committed to the increase and diffusion of knowledge about the world and its life forms. Conducting such a survey and publishing its findings would materially further that goal.

The officers from Deseret were persuaded by this reasoning, and therefore requested that someone at the Smithsonian prepare a formal proposal outlining the nature and scope of the work. Kellogg later met with Philip S. Humphrey, curator of the Smithsonian's Division of Birds, and set him the task of writing up a research proposal for submission to the US Army Chemical Corps Biological Laboratories at Fort Detrick, some fifty miles to the northwest, in Frederick, Maryland.

By October 2, 1962, Humphrey had drafted the six-page "Proposal of Research to be undertaken by the Division of Birds, Smithsonian Institution." This was the foundational document of the Smithsonian's Pacific Ocean Biological Survey Program (POBSP). It laid out the basic set of activities that multiple teams of Smithsonian scientists were to be engaged in for at least the next two years, beginning on October 15, 1962, and ending on October 14, 1964.

Humphrey divided the research into two main fields of concentration, the first of which was an "analysis of bird migration in the Pacific area." What was meant by "the Pacific area" was at this point left undefined so that it potentially included any expanse of open sea in the entire Pacific Ocean. That vast expanse of water was known to be home to four types of birds: those that tended to remain on a specific island; those that moved from island to island on a limited basis; migrants from areas outside the Pacific; and "pelagic forms," that is, birds that lived mostly on or above the open sea and came to islands primarily for breeding and nesting.

"At the present time little is known of migratory movements of birds in the Pacific area," Humphrey wrote in his proposal. "Information on this subject has not been collated or analyzed; at present, it is not possible to describe seasonal movements of all birds of the whole Pacific area in terms of dates of arrival and departure, abundance, and source of migrants. The picture is complicated by the fact that certain birds nesting on islands near the equator do not migrate in relation to an annual cycle."

To make a start, Humphrey suggested that first a literature search be done, along with an analysis of bird banding records held by the US Fish and Wildlife Service. Here Humphrey proposed using processing technology to analyze the information: "The data analysis based on literature and specimens will depend for its effectiveness on a specially printed Royal-McBee data processing card and appropriate sorting equipment."

But to obtain the most current data on bird migration patterns, it was necessary to perform a new banding study. This would require a program of fieldwork in which many species of birds would be banded on a mass scale. "It must be remembered," Humphrey wrote, "that banding provides the only *proof* that an individual bird has moved from one place to another."

He also proposed marking certain birds with colors. "Color marking of pelagic species might also be possible. If a bird look-out could be kept on naval vessels operating in the Pacific much additional data might be obtained."

The second area of concentration Humphrey defined was a "study of the ecology of birds and mammals on one or more Pacific islands." Once again, no particular island or group of them was identified by name; in fact, not even so much as a general region within the Pacific was singled out for study. That, apparently, would be left for the army to decide, or reveal, later. Finally, the proposal included a mandate for taking blood samples and for capturing live birds and mammals.

"Blood samples and ectoparasites will be collected from many of these specimens by a Navy corpsman assigned to the field team," Humphrey wrote. "Live birds and mammals will be trapped for banding and collection of blood and ectoparasites. Some individuals will be kept in captivity for transport to Conus," he said, using the military acronym for "contiguous United States."

"It is to be expected," Humphrey added, "that details of blood sampling requirements will be provided to the project." Provided, that is, by the army. Even at the proposal stage, therefore, the Pacific Project explicitly included significant roles for army and navy personnel.

The proposal concluded with a breakdown of facilities needed for the operation, a list of personnel and their salaries, and a catalog of required equipment and supplies, including, improbably, a "harpoon gun" and four "16 ga. double-barreled shotguns." Shotguns were part of the dark underside of ornithology, for they were the means by which many bird specimens were obtained for study, skinning, and display. Indeed, Charles Darwin himself is said to have collected some three hundred bird skins, shooting down the specimens with a shotgun, while John James Audubon is thought to have killed as many as four hundred birds in order to depict them accurately in his paintings.

Humphrey named himself the project's principal investigator, which he would in fact become. This was a logical choice, reflected by the mini-CV he provided for himself, which described a long string of academic distinctions, fellowships, awards, and prizes. From 1957 to 1962, just prior to coming to the Smithsonian, Humphrey had been assistant curator of ornithology at the Peabody Museum of Natural History at Yale. Coincidentally, the museum's director at that time was S. Dillon Ripley, who would later become secretary of the Smithsonian. When Humphrey himself moved to the Smithsonian in 1962, he would become curator of birds at the National Museum of Natural History. And, although his mini-CV did not mention it, he was also a US Air Force veteran.

* * *

On October 2, 1962, Philip Humphrey wrote up a cover letter in the form of a memorandum to Leonard Carmichael, then secretary of the Smithsonian Institution. Carmichael was an unusual leader for the institution for a couple of reasons. He was the seventh secretary in the Smithsonian's history and the first to be hired from outside rather than promoted from within. He therefore brought with him some new ideas and perspectives, and he began a major program of modernization and expansion. He pushed for new research programs, fieldwork, publication, and exhibits. Predictably, Carmichael had an illustrious set of credentials, with a PhD from Harvard, where he pursued a special concentration in the determinants of human and animal behavior. He had previously taught at Princeton, Brown, and Tufts Universities, and in 1938 he became the president of Tufts.

The most surprising element of Carmichael's background, however, was his connection to the Central Intelligence Agency and its notorious Project MKULTRA, whose purpose was to discover, test, produce, and stockpile chemical and biological materials capable of producing human physiological and behavioral changes. In other words, instruments of mind control. A component of MKULTRA was an entity known variously as the Human

Ecology Fund, or the Society for the Investigation of Human Ecology, or the Human Ecology Society. Whatever its actual name, this was a CIA-controlled funding mechanism for studies and experiments in the behavioral sciences. According to research by the investigative journalist Ted Gup published by the *Washington Post* in 1985, Carmichael was a director of the Human Ecology Fund between 1959 and 1963; he had also "signed a secrecy agreement not to disclose its CIA funding." Thus, he was no stranger to clandestine operations.

However implausible all of it was, this would not be the Smithsonian's only association with the world of espionage: Carmichael's successor, S. Dillon Ripley, had worked as a spy for the Office of Strategic Services (OSS). In any case, Carmichael had no objection to the Smithsonian accepting funds from the US Army Chemical Corps. In his cover letter to him, Humphrey explained the purpose of the attached six-page proposal and then said at the end: "Officials of the sponsoring agency are eager to have the final and official proposal in hand as soon as possible so that the project can begin on October 15; they urge that the fiscal arrangements for this project be made on a contract reimbursable basis."

Humphrey transmitted the proposal and cover letter to Carmichael through Remington Kellogg, who scrawled at the bottom of the letter: "Urgent. This afternoon RK."

A week later, on October 9, 1962, Leonard Carmichael signed a fifty-two-page contract, "Studies on the Distribution, Ecology, and Migration of Pacific Birds and Mammals," with the US Army Biological Laboratories, Fort Detrick, to conduct a two-year program of field research in the Pacific, in return for a total payment of $208,000 (see appendix I). This was the formal and official start of the Pacific Project. By the terms of the agreement, the Smithsonian Institution became a literal contractor to the biological warfare division of the US Army Chemical Corps, the branch of the army charged with producing weapons of mass destruction.

The original "negotiated contract," which is available in its entirety at the Smithsonian Institution Archives, sets forth the respective obligations of the two parties: the Smithsonian Institution ("the Contractor") and the US Army Biological Laboratories, Fort Detrick ("the Government"). Particularly noteworthy was a provision concerning clearance for publication, which stated in part, "No information developed in the performance of this contract shall be published or divulged in any thesis, writing, public lecture, patent application and the like without the prior submission of the manuscript or material to the *Contracting Officer, Fort Detrick*, or his duly authorized representative for clearance. The Contractor agrees to be bound by the decision of said official." That was not by itself a prior restraint on publication, although it was arguably an assertion that, in principle, such prior restraint could be exercised by the government at its discretion. These were the only strings attached to the Smithsonian's participation in the proposed project.

The contract itself bore no security classification on any page; that is, it was not itself marked Secret, Classified, or even (U), for Unclassified. The factual relationship between the two parties, therefore, does not appear to have been secret. Still, there was a further provision regarding technical reports, which were formal reports on the project intended specifically for the army, as opposed to scientific papers intended for open publication to the general public. Some of these technical reports would be classified and, the contract said, may be obtained only by "Qualified requesters." Otherwise, "the transmission or the revelation of its contents in any manner to an unauthorized person is prohibited by law." "All technical reports," the contract said, "shall be transmitted directly to: Commanding Officer, U.S. Army Biological Laboratories," at Fort Detrick.

What is not stated in any part of the POBSP contract, or in any surviving correspondence, memoranda, or other document pertaining to the agreement, is *why* the army was interested in collecting the vast amount of data the Smithsonian's field teams would end up generating. Nor is there anything in the public record to show that a Smithsonian official ever asked what the army intended to do with the data or what the overall purpose of the project was. It's as if it didn't matter. The Smithsonian's attitude seemed to be, we are providing information; what the army does with it is its business.

* * *

If there's anything more incongruous than the Smithsonian's subcontracting itself to the American biological warfare project, it's the story of how the place itself came into existence. The Smithsonian, after all, is the quintessential American institution. Located in the nation's capital, it's one of the world's largest, most popular, and best-known museums. Further, it constitutes the biggest single presence on the Mall in Washington, its several branches and divisions arrayed in two long, parallel rows of buildings facing each other across the grassy central promenade. Anyone would naively imagine, therefore, that the institution was founded by an illustrious American such as George Washington, who early on in life was a surveyor, and who in his final address as president urged that the nation "promote, as an object of primary importance, institutions for the general diffusion of knowledge," a sentiment that captures the essence of the Smithsonian to a T. As founder of the place, Washington would have been symbolically correct and thematically perfect.

But George Washington was not the father of the Smithsonian Institution.

Then perhaps it was Thomas Jefferson, who was himself a natural historian of some repute? Or maybe even Benjamin Franklin, who started the nation's first scientific association, the American Philosophical Society, in 1743? Or at the very least, it must have been founded by someone named Smith. But in fact the Smithsonian Institution was not founded by an American at all, famous or otherwise.

The place owes its existence to a little-known chemist and mineralogist born illegitimately in Paris under the name Jacques-Louis Macie. Macie lived his entire life in Europe, never once set foot in America, and died in Genoa in 1829, under the name James Smithson.

A few years before his death, Smithson made out a will leaving a large portion of his considerable estate to his nephew, Henry Hungerford, and the remainder to Hungerford's heirs. But in his bequest Smithson presciently included a contingency clause in case Hungerford died intestate, or without an heir: "I then bequeath the whole of my property . . . to the United States of America, to found at Washington, under the name of the Smithsonian Institution, an Establishment for the increase & diffusion of knowledge among men."

Now this was an odd situation. The initial legatee, Henry Hungerford, was a high-living dandy who died in a hotel in Pisa in 1835, at the age of twenty-eight or twenty-nine. He never married and had no children, and Smithson's contingency clause came into effect after all. And so the US government was suddenly in for a surprise gift valued at $500,000 in the currency of the time. (According to the Bureau of Labor Statistics consumer prince index, that is equivalent to more than $15 million today.)

But as it turned out, the US government did not want Smithson's money, at least not at first. Members of Congress raised a long series of questions about its source, as well as several objections to accepting it. Who was this mysterious James Smithson anyway, and why was he giving this gift to America? Was this an unknown Britisher with delusions of grandeur who was trying to purchase immortality with an infusion of cold cash? And if we took the money, what exactly should we do with it? Is accepting such a gift even constitutional? There were no clear or obvious answers to those questions.

A few congressmen proposed that the government bluntly refuse the bequest. After it was finally accepted, others argued that the money should be returned. In view of this tortuous debate, uncertainty, and irresolution, prolonged beyond all reason, it is fair to say that the Smithsonian Institution *almost* did not come into existence. But it did, and the series of events that culminated in its eventual founding, in 1846, fully seventeen years after James Smithson's death, makes for a Byzantine story line, one that at times bears an all-too-close resemblance to a US sitcom, a comic opera, or a French farce.

* * *

The story begins in England, with Hugh Smithson, who as the Duke of Northumberland was one of the richest men in the country. Indeed, all the principal figures surrounding James Smithson were fairly wallowing in property and cash. In 1740, Hugh Smithson married Elizabeth Seymour, of the socially prominent Percy family. Seymour was even richer than her husband at the time of their marriage, and in 1745, on the death of her father, Hugh Smithson took the name of Percy, thus becoming Hugh Percy. (As an aid to confusion, not only Hugh Smithson but also James Smithson himself and other members of the family circle had a bad habit of changing their names, sometimes more than once.)

The establishing event in the narrative arc of James Smithson's life occurred in 1764, when Hugh Percy had an affair with *another* Elizabeth, Elizabeth Hungerford Keate Macie. *This* Elizabeth had so noble an ancestry that in the far upper reaches of her family tree appears the imposing figure of Henry VIII, famous beheader of wives. During the affair with Hugh (Smithson) Percy, Elizabeth Macie became pregnant, and to conceal her pregnancy and give birth in secrecy, she left England for Paris. The child, a boy, was now given the first of his many names: Jacques-Louis Macie. He is thought to have been born in 1765.

Jacques-Louis Macie spent his early years in France, and French was his native language. In 1774, however, Elizabeth brought him to England, where under the Anglicized name of James Louis Macie he became a British citizen. At the age of seventeen he entered Pembroke College, Oxford, where he enrolled under the Latin version of his name, now styling himself Jacobus Ludovicus Macie (later changed to James *Lewis* Macie). In 1786, at the age of twenty-one, he received a master of arts degree.

By that time the main interests of Macie's life were clear: they ranged over natural history in general, and in particular the sciences of chemistry and mineralogy. He was a young man who wanted to know what there was to be known about the natural world and its makings. To that end, he visited Fingal's Cave in the Hebrides Islands of Scotland (the setting of an overture by the composer Felix Mendelssohn), from which he took scrapings of the black basaltic columns that are a geological feature of the cave. The excursion to and into the cave was a hazardous one, that took considerable courage on James's part to undertake.

In 1787, James Macie attended a meeting of the Royal Society, England's premier scientific association, discussion group, and social club. It existed to promote "natural" knowledge as distinct from superstition, dogma, and supernaturalism. James became a full member, which meant that he was now traveling among a sophisticated set of scientists, including Joseph Banks, who was then the society's president, as well as chemists Henry Cavendish and Humphry Davy, and others of comparable reputation and accomplishment.

Over the course of his career as an itinerant geologist, Macie collected some ten thousand mineral specimens and published more than two dozen scientific papers. If he had any claim to fame among chemists, it was in showing that what had long been regarded as a single substance, calamine, in fact existed in several distinct varieties, two of which were zinc silicate and zinc carbonate. In honor of his discovery, chemists renamed the latter compound "smithsonite," a term that is still in occasional use today.

In 1800, James's mother died, leaving most of her property to her first son, who at that point was a decidedly wealthy man. Soon after her death, and perhaps in compliance with a prior request of his mother, James adopted his father's original last name, although it took an act of Parliament to make the change official. But in February 1801, at the age of thirty-five or thirty-six, he at long last became James Smithson.

It was also in 1800 that Smithson helped create the Royal Institution, which was and is separate from the Royal Society. It was founded to bring science to the masses, and some have speculated that the Royal Institution was the model for Smithson's bequest to the United States. But why the United States rather than his own country, England? Was it because England was already well-endowed with scientific institutions and societies, whereas the young nation was not? The answer is unknown. It is known, however, that in the late 1810s, Smithson had a falling out with the Royal Society, apparently over what he regarded as the editorial mistreatment of one of his papers.

In 1826, Smithson wrote out a will, using as his guide the booklet, *Plain Advice to the Public to Facilitate the Making of their Own Wills*. At length

Smithson moved to Paris, where he indulged a newly discovered passion for gambling. And then, in 1828, in poor health, he packed up his belongings and moved south, to the Mediterranean. He wound up in Genoa, the birthplace of Christopher Columbus.

It was there that in the early morning hours of June 26, 1829, James Smithson died.

Many years would pass before anyone in the United States learned that a relatively minor British chemist had left a fortune to the US government. The reason for this is that Smithson's will specified that the immediate beneficiary was his nephew, Henry Hungerford. Like everyone else involved in the Smithson saga, the nephew underwent his own course of name changes. At the end of them, Hungerford adopted what he evidently regarded as the stylish and sophisticated last name of his stepfather, Theodore de la Batut. And, in a final act of faux elegance, he also titled himself "Baron" in the process. Thus, Henry Hungerford became Baron Henry de la Batut.

After receiving his inheritance from James Smithson, Baron de la Batut spent the next five years frittering it away in Europe, "living for his pleasures," as it was said, before dying, without heir, in Pisa in 1835. Thus James Smithson's original contingency clause leaving the residue of his funds to the United States took effect.

And so it came about that, in the words of Smithson biographer Heather Ewing, "The death of this obscure young man set in motion an extraordinary sequence of events, which led ultimately to the birth of the Smithsonian Institution."

That extraordinary sequence of events, however, can be told in short compass.

* * *

News of Baron de la Batut's death traveled to London, specifically to James Smithson's solicitors, who in turn notified the US chargé d'affaires in London, Aaron Vail, of Smithson's bequest. On July 28, 1835, Vail wrote a letter to John Forsyth, the US secretary of state, informing him of "the particulars of a bequest of property to a large amount left to the United States by a Mr. James Smithson, for the purpose, as stated in the will, of founding in Washington an institution 'for the increase and diffusion of knowledge among men.'"

Secretary of State Forsyth was thus the first person in the United States to receive word of Smithson's gift. He promptly sent the news to President Andrew Jackson, who in turn notified Congress. Jackson claimed that he himself lacked the authority to accept or reject the money and that it was for Congress to decide.

Few in Congress were amenable to the idea of accepting Smithson's money. However, John Quincy Adams, the former president and now congressman, was wholly in favor and persuaded the Congress to send someone to London to at least take delivery of the funds.

That process in itself took two years because of a backlog of cases in the British Court of Chancery. But finally everything was sorted out, and Smithson's bequest, in the form of 104,960 gold sovereigns packed in 105 leather bags, was dispatched to the United States. After its arrival on US shores a month later, the money was deposited into the US Mint in Philadelphia.

Now began a battle that would last for eight years, be carried out in the form of 458 congressional discussions, the rhetoric of which would take up four hundred pages in the Congressional Record. In the end, an agreement was reached to create a "Smithsonian Establishment." It would consist of a museum, library, art gallery, and miscellaneous other such establishments.

And so, belatedly, on August 10, 1846, eleven years after the bequest reached the United States, and seventeen years after the death of James Smithson, President James K. Polk signed into law the bill that would create the Smithsonian Institution.

Had he known the degree of travail and controversy that his bequest would cause in the United States, Smithson might well have offered his money elsewhere. But as it turned out, he would be responsible for the creation of one of the world's greatest museums and one of the most familiar names in the annals of science history, research, and public education.

2

Recruitment

By the time the Pacific Project was getting underway in the 1960s, the Smithsonian had become a beloved, almost hallowed—indeed practically revered—national institution. After its founding, it was led by a succession of "secretaries" that included scientists such as the physicist Joseph Henry (the *henry*, the unit of electrical inductance, was named after him) and the aviation pioneer Samuel Pierpont Langley. Its various buildings on the Mall in Washington held well over a million artifacts, including bird and animal skins, dinosaur skeletons, and iconic national treasures such as the *Spirit of St. Louis* and the original Wright *Flyer*. The institution had acquired an international reputation for its high standards, probity, and an air of purity and good judgment in its collections and in its general approach to matters of science.

But despite its spotless reputation, Philip Humphrey, director of the Pacific Ocean Biological Survey Program (POBSP), did not have an easy time of securing recruits for its fieldwork. The ornithological community tended to be a small and closely knit group, and news of the project spread not by means of written announcements or advertisements placed in journals but rather by word of mouth.

"Don't know if there was an announcement or advertisement of the project," said Fred C. Sibley, one of the first field team members to be hired. "I heard about it from a friend working at the Smithsonian."

"I don't recall formal announcements," said Charles Ely, who would become field director and head of the program's Honolulu office. "There were a lot of inquiries at meetings, especially from individuals who had actually experienced collecting and often difficult field work."

The program's military funding turned off many professors of ornithology, who worked hard to dissuade their students from accepting positions with the POBSP, despite the relatively high salaries being offered.

"Humphrey was scrambling to find people and struggling to get more information from the Army contractors," Fred Sibley recalled. "The ornithological community was suspicious and advising their students not to apply—my major professor being no exception." The problem was the army connection. "Professors were telling their students to steer clear of this military contract."

Beyond the issue of army funding, the original two-year duration of the contract was also a deterrent to those looking for more permanent, tenure-track positions in colleges and universities. Nevertheless, over a period of months, Humphrey gathered an initial band of field trip personnel. They were all men; the few women in the program were restricted to secretarial, administrative, and laboratory work. Maryanna Smellow, for example, was a research secretary in the Smithsonian's Division of Birds, and Jane P. Church was officially the project secretary, but in reality was much more than that. Some have called her "the heart and soul of the POBSP," while the fieldworkers themselves regarded her as their "den mother."

As of March 1, 1964, Jane Pickens Church started off with the title of "laboratory aid" but was soon made an administrative assistant to the project director, Philip Humphrey. According to Max Thompson, one of the field team members who knew her well, "She ran the day-to-day operations in Washington and was a link from Washington to those in the field. We usually dealt directly with her if we needed something. She was the liaison between the Bird Banding Office and the project. She usually had two to three secretaries working for her. She also was frequently in contact with Deseret and John Bushman [military liaison]. She was an excellent organizer.

"As for being a 'den mother,'" Thomson continued, "she certainly looked after us to make sure we had what we needed. She would often take us to her home in Leesburg for the weekend. Her folks owned 360 acres on the Potomac and lived in what most of us would call a mansion. She lived with her parents, the Pickens, who were well known in the area."

The first field team members to be hired ranged from trained researchers with advanced degrees in ornithology to college dropouts and included others with no professional ornithological background at all. There was an artist, Allen Blagden, and a jazz musician, Ray Jillson. "He had a pet blue jay that he gave to the zoo eventually," Roger Clapp, another project member, recalled of Jillson. "He did not have professional training so far as I know, he just liked birds. Heard about the project through the grapevine and was accepted into it."

There were even two high school students, Peter Marshall and Mike Trevor, whose participation in the program amounted to summer jobs.

Fred Sibley, in contrast, was one of the most well educated of them all. To start with, both of his parents were themselves scientists. His father had a PhD in entomology, his mother a degree in botany, both from Cornell, where they met. "My mother," Sibley recalled, "was convinced that Ithaca, New York, was the center of the universe."

It was the center of the ornithological universe, anyway, as one of the world's top bird experts taught there: Charles G. Sibley, who was no relation to Fred. (But David Sibley, author and illustrator of the well-known *Sibley Guide to Birds*, was and is Fred's son; the two regard ornithology as "the family business.") It seems that Fred was preordained to be a part of the POBSP and to be a leader of many of the field trips. "When I was 5 or 6 a relative asked me what I wanted to do and I said, 'Lead expeditions.' The child evidently knew my future."

Fred got his bachelor's degree from Cornell in 1955, and a master's in 1959, with a thesis on the hybridization of towhees (songbirds of the bunting family) on the Mexican Plateau. It was while leading an expedition to Mexico as a grad student that Sibley became friendly with Roger Clapp, another Cornell birdman who would join the POBSP. Later, Sibley became a lecturer in biology at a branch of University College London in Ibadan, Nigeria, where he taught for two years. He returned to the States, and in the fall of 1962 got a job teaching comparative anatomy at Adelphi University, on Long Island.

But given that he had a wife and four kids to support, Adelphi's salary proved too low for comfort. "All the faculty were dependent on the extra income from evening classes," Sibley recalled. "Because of a drop in evening enrollments and my being the newest faculty member, there were no courses I could teach to raise my salary ($6,000) to a living level."

One of his colleagues suggested that he join the newly created Pacific Project. "Since this would be a 50% salary increase, and later 100%, I jumped at the chance. Phil Humphrey, in charge of the proposed project, called me October 17, 1962, and I committed to joining at the end of the semester (late January)." That was fast work on Humphrey's part: the contract had only been signed on October 9. This made him one of the earliest recruits into the POBSP.

Another was Roger B. Clapp. If anyone was destined by his DNA, the stars, and fate, to be part of the POBSP, it was he. Clapp's interest in birds had developed early in his life, starting at about age five. He and his father were on a hike near their home in Amesville, Connecticut, when a bright red bird streaked across Roger's field of vision. He immediately wanted to know what it was, but his father was by no means a bird fanatic. (The bird proved to be a male scarlet tanager.)

Later, Clapp's grandmother gave him a birding book, one of the Chester A. Reed pocket guides. But in Clapp's view, the Reed books were hardly the best of the genre. "These guides provided some truly vile illustrations of birds and some descriptions," he recalled, "so that my ambitions were put on hold for some time."

A few years later he finally came across a true and proper bird identification manual: *Peterson's Field Guide to Eastern Birds*. He traded his entire stash of comic books—some eighty in number—to acquire a copy from a friend. After this, making a positive identification of even a relatively unusual specimen such as a pine grosbeak became an easy matter. He also became adept at identifying a wild bird merely by its song.

Most of Roger's friends at the time were geeks interested in cars and rockets, but he wasn't. Indeed, Roger Clapp gradually became so consumed by bird watching that in high school he started taking Fridays off just to go birding, accompanied by his dog, Rex, or by his cat, Sam. Still, he was getting straight As in his classes and wound up as the school's very first National Merit Scholar.

Clapp applied to only one college, Cornell, which he knew had an ornithology program, run by Charles G. Sibley. Sibley was a man of strong opinions who was not well-liked by some of his colleagues, and who in turn did not suffer fools lightly. He also hated to be one-upped in the field, which Clapp had a habit of doing, calling out bird identifications usually a split-second before his professor, the expert, managed to.

One of Charles Sibley's research projects was to study the proteins of bird egg whites to better establish relationships among the higher taxonomic categories of birds. For this purpose he needed lots of specimens and in 1958, Sibley sponsored a field trip to Mexico to collect the eggs of various species.

It was on the trip to Mexico that Roger Clapp met Fred Sibley. Clapp realized from this experience that he was easily bored by what he called "dooryard birds," the kind that anyone could find in their backyard: bluebirds, robins, sparrows, crows, and the like. What excited him was spying his first wild parrot, and more exotic birds such as motmots, trogons, jacanas, and the astonishing variety of hummingbirds found in the Mexican lowlands.

He also decided during the expedition that he really loved fieldwork (especially as opposed to attending classes). But he also discovered that living

in the wilds was not without its dangers. One night, as he and Fred were preparing bird skins in their tent, they both heard a loud popping sound.

"Firecrackers?" Roger blurted out. But Fred was grabbing at his leg, hit by
a .22-caliber round fired by an unknown shooter, for unknown reasons. Fred
was taken to a Mexican doctor who refused to remove the bullet in Fred's
leg on the grounds that he, the doctor, still had a bullet in his chest from
the Mexican Revolution. (A month later, back in Ithaca, another physician
removed the bullet.) Roger, meanwhile, spent the night alone in the tent,
which now had a bullet hole in the wall near where he had been sitting.

"That humid night, with hundreds of enormous fireflies drawing large J's
through the air, is certainly one that I never will forget," Clapp reflected later.

At length Roger Clapp became an expert, practical, hands-on ornithologist. He learned how to skin and mount birds, find nests, and collect eggs,
which was an art in itself. He was unusually committed to a task or challenge,
whatever it was. On one occasion he spotted a black tern nest floating on the
waters of a marsh in Ithaca. Clapp swam out to collect the eggs and brought
them back in his mouth.

Shortly after graduating from Cornell, Roger heard about the Smithsonian's
Pacific Project. In June 1963, he wrote Philip Humphrey to apply. In an interview at the Smithsonian, Humphrey told him about the job, which was to
make a systematic survey of the life forms on some islands in the Central
Pacific. The survey was not limited to birds: its purpose was to find and inventory virtually everything that lived on these islands: birds, mammals,
reptiles, parasites, vascular plants, "a full biological survey."

Humphrey said nothing about who was funding the project, and Clapp
didn't ask. "Why would I care?" he said later. Anyway, he got the job. He
started out at the bottom, as a technician, but was soon promoted to research
curator, the salary for which, during the first year of employment, would be
$7,100.

* * *

Many of the findings and results of the Smithsonian's surveys were published
in various scholarly journals, often in the *Atoll Research Bulletin* issued by
the Smithsonian. Some of these studies were extremely detailed and physi-
cally immense. *The natural history of French Frigate Shoals*, for example, ran
to 383 pages of text. It was written by a single author, A. Binion Amerson Jr.,
one of the most prolific chroniclers, as well as perhaps the longest-serving
member, of the Pacific Project. The length of Amerson's report on French

Frigate Shoals is surprising in view of the fact that the total land area of its thirteen named islands was only 0.09 square miles. The report's comprehensiveness is a testimony to the care with which Amerson researched the history of the islands and collated and tabulated the large number of data points produced by the fieldworkers.

Amerson was born in Macon, Georgia, in 1936. His grandparents were farmers, his mother was a gardener, and early on the boy developed a love for the outdoors. He was also interested in science, however—mainly biology, physics, and chemistry—and he had a knack for thinking up somewhat odd-ball scientific experiments that he hoped would prove something or other. For example, Binion wanted to know what the effect of x-rays might be on plant growth. In an effort to find out he took some seeds from the petunias that his mother grew and brought them to the family dentist, who duly irradiated them.

"There was some minor interference to the growth of the plants," he remembered years later. "Not much of an effect." Nevertheless, Amerson wrote up his results for his high school science fair. As a senior, he put together an electric piano after reading an article in *Popular Mechanics* about how to use an electronic device to produce a musical tone.

During the summers Binion worked for the Department of Agriculture in Georgia and learned about plant pests—Japanese beetles, ticks, mites, and the like—and developed an interest in such things. So when he entered Mercer University in Georgia he made a beeline, as it were, for the entomology department. He got his bachelor's degree in biology there, with minors in physics and chemistry.

Later, Amerson enrolled in grad school at the University of Kansas. KU was something of a mecca for ornithologists. It boasted its own Natural History Museum, whose first curator of ornithology was none other than the peripatetic Charles G. Sibley. Later, Philip Humphrey would become the museum's director. Max Thompson did graduate work there on the birds of North Borneo.

While at Kansas, Binion Amerson took courses from J. Knox Jones Jr., a mammalogist. In the spring of 1962, the US Army Medical Research and Development Command awarded Jones and others at the university a contract to perform an ecological study in Mexico. The rationale for this adventure, Binion was told, was that the army wanted to prepare for a possible invasion of the North American mainland by Fidel Castro via the Yucatán Peninsula, from which point he might eventually infiltrate the United States.

The army had scant knowledge of the wildlife of the Yucatán and wanted to know the kinds of animals, poisonous reptiles, and insects US soldiers might face. This was after the Bay of Pigs invasion (April 1961) but before the Cuban Missile Crisis (October 1962).

Amerson joined the project, which was a classified military operation. At this point, then, he was already doing secret undercover work for the army, so that when another project also funded by the army, the POBSP, appeared on the horizon, he had no qualms about it.

His time in Mexico was also his introduction to fieldwork in the tropics, as he and four others collected a wide variety of specimens. They captured bats overnight in "mist nets," lengths of open-mesh material stretched about twenty feet between two poles or trees with pockets at the bottom to collect bats snared by the nets. During the day the team members captured birds in the same manner. They also set traps for snakes and other reptiles, killed and skinned them, and brought them back to the lab in Kansas for later study. They ran combs through the feathers of live birds, looking for fleas, ticks, and other parasites.

One night he attended a concert in the Great Ball Court at Chichén Itzá, site of the Mayan ruins in the Yucatán. A hush passed over the audience during intermission as people pointed to the sky.

"It was Sputnik," Amerson recalled (though wrongly: *Sputnik* was in 1957; this had to have been a later Russian satellite, *Vostok 3* or *4*). "It was an eerie feeling," he said, "because here we were in this archaic, ancient setting, the Mayan Ball Court, and here comes this modern Russian spaceship!"

Altogether, his experience in Mexico constituted an ideal preparation for the Pacific Project. So much so that in Amerson's case, the army called him, not the other way around.

"The call came in at night," he remembered. It was from K. C. Emerson, of the Office of the Assistant Secretary of the Army, calling from the Pentagon. Emerson was himself an entomologist and later wrote a book wittily entitled *Lice in My Life*. Emerson knew of Binion's work for the army in Mexico and asked whether he would be interested in joining a similar project to be conducted by the Smithsonian. It was in the Pacific, and it involved collecting birds, mammals, and parasitic insects on various uninhabited tropical islands.

"He said the project was to be funded by the Defense Department," Amerson said. "Didn't say why."

But it made no difference to him. Amerson, who was at that point still in his twenties, thought it would be a fine way of spending the next two years of his life, even if it meant dropping out of grad school for a while. And so in a matter of days, he was aboard a flight to Washington. He was among the first group of fieldworkers hired, and he stayed with the project until the end.

* * *

In all, Philip Humphrey and the Smithsonian hired some sixty-five staff members for the Pacific Project. One of the youngest—likeable and well educated but somewhat naive in his way—was Larry Huber.

"He unwittingly tended to do the socially incorrect things," Fred Sibley said. "Thus the multitude of Huber stories."

One concerned his dexterity, or the lack thereof, with a bowie knife he carried around with him from time to time in the wilds. In 1964, during one of their visits to an inhabited island, the POBSP field team members much admired how quickly the Marshallese natives were able to husk coconuts, with just a few fell strokes of a machete. Huber offered to demonstrate his mastery of their technique, but the knife slipped, and he wound up stabbing himself in the chest, fortunately only nicking his breastbone.

Huber was born in 1944 and raised mostly by his mother. Larry spent his early years in Maumee, Ohio, where he went to high school. He was interested in wildlife and collected everything from bugs and butterflies to snakes and turtles. At one point he took a correspondence course in taxidermy and learned how to mount specimens of birds and mammals, most of which were roadkill. He later attended the University of Arizona, where he received a BS with a major in ornithology and a minor in herpetology.

Larry kept a small collection of reptiles in his dorm room, some of which were venomous, including live snakes and a Gila monster. He was once bitten by one of his own rattlesnakes, and on another occasion also bitten by the Gila monster, but he survived both experiences.

When one of his professors heard about the Smithsonian's Pacific Project, he suggested that Larry apply for a position, which he was happy to do. Huber would be one of the POBSP's herpetologists, but it became obvious during field trips that he had decided opinions about other animals. Although he was fond of snakes and lizards, he had an inordinate fear of sharks. He loved birds but had a hatred of cats, who killed birds to the point of decimating avian populations on some of the islands they surveyed. He therefore made a specialty of hunting the cats at night and killing as many as he could, which was a lot.

As a result of his early acquaintanceship with taxidermy, Huber was adept at preparing and mounting bird skins. He also had a novel idea for constructing nesting sites for red-tailed tropicbirds, a particularly graceful white bird species whose members sported long, thin, exceptionally brilliant red tails. The tropicbirds liked to nest in structures such as old houses, under downed tree limbs, in caves, or below rocky overhangs.

On Jarvis Island, near the equator, Huber came up with the idea of constructing artificial nesting sites for the tropicbirds by placing slabs of coral rubble in an inverted V formation, thereby creating a shelter. These structures were immediately and widely used by the birds. Photographs of the shelters in the Smithsonian archives show them still occupied by tropicbirds some eight to ten years later.

Soon after their hiring, all these men—Fred Sibley, Roger Clapp, Binion Amerson, and Larry Huber—were on their way to the Pacific. Sometimes together in various combinations, other times apart, all four set out on what would turn out to be, in their own view, the grandest and greatest experience of their lives.

* * *

Once hired, the POBSP recruits faced a series of tasks and requirements prior to their departure for Hawaii. In 1964, the project leaders established a separate headquarters in Honolulu, in a house provided for Charles Ely, the project's field director, and his wife, Janice. On holidays, the two would sometimes host gatherings for field team members who happened to be in town. Honolulu would therefore be the first stop for the field teams, an intermediate staging point before the men were sent out into the open ocean and, finally, to the islands. Being a joint operation run by two big organizations— the Smithsonian and the US government—a certain amount of red tape was inevitable.

First was the matter of security clearances. Because the fieldworkers would have access to background information pertaining to the military program, and because that information might include items officially classified as Secret, each individual had to be issued a "facility security clearance required for contract performance."

Supplying the clearances was the province of the Office of Naval Research in Washington, and the process involved the submission of forms pertaining to each of the fieldworkers. The clearances, when obtained, would be distributed to officials at agencies including the US Army Biological Laboratories at Fort Detrick, the Deseret Test Center, and the director of the FBI.

What was and what was not to be regarded as "classified" was somewhat hard to pin down, and a year into the project the whole matter of secrecy and security became a touchy subject. An April 20, 1964, letter from Charles Ely to the project's research curators started off: "As the result of a recent security meeting in Washington some aspects of our program have been classified by the military. It therefore becomes important that our people be even more careful in discussing the project with outside people. No one wants to be branded a security risk as the result of an idle conversation.

"Since our contract is from the Department of Defense," the letter continued, "we are under routine security regulations and discussion of any classified material with non-appropriate parties can be a serious offense. Our ecological studies are a normal S. I. type of research and are unclassified. When our activities become associated with other groups however, they *may* become classified. Items which are unclassified today could be classified tomorrow as we have discovered."

Ely then provided a list of things that field team members should be careful about when talking with outsiders. One of them was, "Don't use any naval ship names with the term S [Smithsonian] or with geographical location or with any non-S.I. part of the project." Another was, "Don't associate DTC with S." And, "Don't mention the eastern organization [Fort Detrick] or live bird shipments in any connection."

Ely then advised his men, "Don't be so secretive that all outsiders think we are on a very top secret operation. Too much secrecy would just add to outside curiosity. . . . Don't scare anyone, our personnel can tell others what we are doing in vague terms. It's our association with other groups that is classified."

The "other groups" were Fort Detrick and the Deseret Test Center. In the end, what all of this added up to was that the normal precision and transparency of science communication was now to be avoided in certain cases. Instead, some issues were to be deliberately muddied, dodged, and carefully obfuscated.

A second task, beyond the security clearances and imperative of secrecy, was getting the men their required immunizations. All the recruits had to make trips to Fort Detrick to be vaccinated for certain diseases before going out into the field. But exactly what those diseases were, and why the inoculations were necessary in the first place, were topics of speculation at the time—and still are. Were they protection against diseases likely to be encountered in the field or against the consequences of contact with pathological agents used in biological weapons trials? Further, how did anyone know what diseases were likely to be encountered on the field trips, given that one of the very purposes of the POBSP survey was precisely to discover what kinds of diseases were already out there?

"We asked what we were being immunized for and told that was information they couldn't divulge," said Max Thompson. "I heard later that one of them was for tularemia. I would like to have known but that wasn't forthcoming."

Roger Clapp had a different recollection. "Rumor was, Q fever," he said. "But a *rumor.*"

There is weak but not conclusive circumstantial evidence implicating both possibilities. Q fever, which was discovered in 1937 (and was named "Q" for query), is caused by the bacterium *Coxiella burnetii.* The pathogen is carried by both wild and domestic animals, birds, and ticks, without the host itself developing the disease. In fact, human beings are the only species that normally contracts the illness as a result of infection. Further, it is thought that a single organism is enough to cause the disease in humans.

But precisely because it was so infectious in humans, and because it caused an incapacitating rather than a lethal disease, the Q fever bacterium was well studied by the US Army as a possible biological warfare agent. Indeed, it was the Q fever microbe that was experimentally disseminated over the group of Seventh-day Adventist human volunteers at Dugway Proving Ground in 1955. In addition, the army kept large stocks of the pathogen on hand at Pine Bluff Arsenal, at least until 1969, when President Richard Nixon ordered the termination of the entire US biological warfare program and the destruction of all its pathogenic biological agents.

Finally, the Q fever pathogen was one of two biological agents that would be disseminated in 1965 by US Air Force jets in an open sea area of the Pacific southwest of Johnston Island. The POBSP regularly visited Johnston Atoll across the years; indeed, one of its islands, Sand Island, had project members stationed on it more or less continuously for the lifetime of the program.

The other agent used in the 1965 field trials near Johnston Island was *Francisella tularensis,* the cause of tularemia. Tularemia is harbored by ticks carried by rodents and is so closely associated with rabbits that another name for the disease is rabbit fever. An effective vaccine was developed by Fort

Detrick scientists Henry Eigelsbach and Cornelia Mitchell Downs in 1961. A Q fever vaccine was also available at the time of the biological weapons tests near Johnston Island in 1965.

Still, the fact that POBSP personnel were stationed at Johnston Atoll would not explain why the project members were vaccinated against the two agents used in the field trials. Pacific Project biologists were not in fact present, and were not expected to be present, at any of the tests in question, which in any case occurred in a grid area some 150 miles to the southwest of the atoll itself. Further, the trials were conducted under tightly controlled and heavily monitored conditions, with all nonparticipant personnel rigorously excluded from the test area. So if the POBSP fieldworkers were in fact immunized against Q fever and tularemia, the reason why remains a mystery.

* * *

Before any actual fieldwork could take place, there was yet one more regulatory hurdle to surmount, and that was to obtain official permission from a few different government agencies for the Smithsonian field teams to actually undertake the work in question. The Pacific is a large, trackless ocean, but some of its islands were United States possessions, and many are and were at the time US wildlife refuges. This meant that for the fieldworkers to capture live birds, or to take blood samples, or to take "etiological specimens," that is, disease-carrying organisms, permission would first have to be obtained from the relevant authorities.

And so as early as October 3, 1962, even before Leonard Carmichael put his signature on the original contract, a certain B. J. Osheroff, the liaison officer at Fort Detrick, had already applied to and received from the Animal Inspection and Quarantine Division of the Department of Agriculture in Washington a formal, written permit to import "ecological specimens during the remainder of the calendar year 1962 and during the calendar year 1963. . . . The ecological specimens may be imported from various Pacific islands for research purposes." According to the permit (No. 1087, "Organisms or Vectors"), the specimens were to be shipped aboard US government aircraft to Honolulu, and for this purpose the agency supplied twelve shipping labels.

A later permit, issued on October 11, 1962, by the US Fish and Wildlife Service, authorized the Army Materiel Command to import into the United States 231 live specimens each of the species golden plovers, wandering tattlers, and ruddy turnstones. "These migratory birds are to be captured on various islands of the West Pacific under supervision of U.S. Army personnel," the permit said, "and are to be shipped in cages or other closed containers by MATS [Military Air Transport Service] aircraft." (A later amendment to the permit advised against taking any bristle thighed curlews, a rare species.)

An additional permit was needed to use "Japanese Mist Nets" or to color-mark birds by dyeing their feathers.

Permission would also have to be obtained even for so noninvasive a procedure as banding. And so in January 1963, Hawaii's Department of Land and Natural Resources issued a Scientific Collecting Permit to Philip Humphrey, Binion Amerson, Robert MacFarlane, and Fred Sibley, allowing them to band "all species" of birds. These men (except for Humphrey) were in fact members of the first POBSP Southern Island Cruise, which set out from Honolulu on February 4, 1963. That same year, these and other individuals were furthermore authorized by the US Department of the Interior to collect "not more than six (6) Hawaiian seals" on the Hawaiian Islands National Wildlife Refuge and also to gather "ectoparasites and blood samples . . . with special emphasis on the collection of data relevant to birds and ectoparasite-borne virus diseases."

In addition to permits, there were also "required procedures to be taken prior to landing on wildlife refuge islands." As the document containing a detailed description of the procedures explained, "The introduction of new plants or insects or animals to one of these islands could, over a period of years, prove disastrous to various bird species which utilize these islands as breeding and nesting grounds. Each island in the chain is, in itself, unique, and this procedure listed below should be followed prior to any and *every* landing made."

The procedures required that before setting foot on an island, all field personnel had to decontaminate themselves and their clothing, equipment, tents, packs, and vehicles, if any. It was a seven-step process that started from the ground up, with #1: "Boots--Remove shoelaces carefully from shoes. If laces are worn, carefully examine for and remove from each lace any burr or seed seen. Spread unlaced shoes wide, and with a pointed object, scrape all seeds, mud, or accumulated dirt from all crevices, particularly where the tongue is sewn into the shoe."

Next was #2: "Trousers." Cuffs had to be rolled down and any accumulated dust or debris brushed away. Pockets had to turned inside out and any foreign material removed.

Further instructions pertained to shirts and jackets, packs, tents and ponchos, boxed equipment, and vehicles, which "should be completely free of debris. This may require washing, dusting of the interior and steam cleaning." (One cannot quite imagine POBSP field team members "steam cleaning" their vehicles, in the doubtful event that they had any to begin with.)

With this draconian set of permits, procedures, and micromanaging dos and don'ts received, recorded, and filed away, the Pacific Ocean Biological Survey Program could finally officially begin.

3

Prequels

The US Army planned to conduct its large-scale biological weapons trials in the Pacific Ocean precisely because it was so vast, so devoid of land, and so sparsely populated, especially by people. The Pacific is by far the world's largest body of water, covering fully a third of the earth's surface, an area larger than all the world's land masses combined.

Some of its islands were settled by the Polynesians as far back in human history as 3000 BCE, thousands of years before any European voyager dared to sail beyond the sight of land. Part of the European reluctance to venture forth was the sheer fear of the unknown. But, as J. C. Beaglehole has noted in his classic book, *The Exploration of the Pacific*, "There comes a time when the terrors of the unknown are overcome by its fascination."

Beaglehole defined three successive waves of European exploration and discovery, and suggested that each had been propelled by a different set of motives. The first wave occurred in the sixteenth century, as a product of the Spanish desire for new sources of wealth. Indeed, it was a search for a faraway land brimming with gold that led the Spanish explorer Vasco Núñez de Balboa to present-day Panama, then called Darien, and to trek across the isthmus. After reaching the western shore, Balboa became the first European to lay eyes the Pacific, on September 25, 1513, an event immortalized by the poet John Keats, in his sonnet, "On First Looking into Chapman's Homer":

> Or like stout Cortez when with eagle eyes
> He Star'd at the Pacific—and with all his men
> Look'd at each other with a wild surmise—
> Silent, upon a peak in Darien.

Historian David Abulafia explains that when Keats was advised that "Balboa, not Cortés, had stood upon that peak, Keats retained the name Cortés, so that the scansion would not be ruined."

After wading into the waters of the South Sea, Balboa claimed it and all of its continents, islands, and whatever riches they contained for his master, the king of Spain. But despite his substantial achievements and his territorial claims, Balboa did not come to a good end. In 1519, six years after discovering the Pacific, Balboa stood trial on the trumped-up charge of usurping the power of a local governor in Panama. He was found guilty, sentenced to death, and beheaded.

The second wave, pioneered by the Dutch in the seventeenth century, was motivated by a simple desire for trade. And the third wave of explorers, represented, for example, by the French Louis Antoine de Bougainville in the eighteenth century, pursued geographical and scientific knowledge for its own sake.

The Pacific Ocean Biological Survey Program (POBSP) fieldworkers were likewise in search of knowledge rather than fortune, and to prepare them for the tasks that lay ahead, a group of Smithsonian researchers led by Gorman Bond, "biologist in charge," performed the literature search that Philip Humphrey had proposed of surveys already done of the Pacific's islands and atolls. He and two assistants duly ransacked the available scientific publications on the birds and other wildlife of the Central Pacific and placed photocopies in reference files.

Independently, it so happened that two of the many prior exploratory voyages into the Pacific had intimate connections to the Smithsonian Institution. The first "prequel" began in 1838 and ended in 1842, four years before the Smithsonian Institution actually came into existence. Indeed, the expedition was responsible for providing the institution with its original and foundational collection of artifacts and specimens. It is not too wild an overstatement to say that, were it not for this expedition, the Smithsonian's first museum, when it finally opened in 1855, would have been somewhat at a loss for items to put on public display.

The voyage in question was the US South Seas Exploring Expedition, informally known as the "Ex. Ex." It was also known as "the Wilkes Expedition," after its commander, Charles Wilkes. (Critics dubbed it "the Deplorable Expedition" on account of the political infighting required to secure authorization and funding.)

The second prequel to the POBSP was the Tanager Expedition of 1923. It was led by Alexander Wetmore, an ornithologist and avian paleontologist who shortly after the end of the voyage would become assistant secretary of the Smithsonian Institution, and, ultimately, its secretary, the immediate predecessor of Leonard Carmichael.

The Ex. Ex. was the larger and more ambitious of the two prior expeditions. According to its chronicler, Nathaniel Philbrick, the Ex. Ex. was "one of the largest voyages of discovery in the history of Western civilization." Further, it had been proposed "at a time when a trip to the Pacific was equivalent to a modern-day trip to the moon."

The idea for the enterprise had been advanced in the 1820s by one Jeremiah N. Reynolds, a civilian science advocate and a fiery, mesmerizing speaker. He used every opportunity to publicize his view that the United States should attempt a voyage that would settle the then–still controversial question of whether there was a continent at the bottom of the world and survey and chart the islands of the South Seas. At length, an upwelling of popular support for this idea among marine and scientific societies eventually reached the halls of Congress, which in 1828 passed a resolution requesting that President John Quincy Adams send a naval vessel into the Pacific.

The voyage that finally resulted was a four-year-long circumnavigation of the globe, undertaken by six square-rigged naval vessels carrying a total of 346 sailors, plus natural historians, scientists, and artists. The expedition set out from Hampton Roads, Virginia, on August 18, 1838, and returned to the States at New York Harbor on June 9, 1842, minus two ships and twenty-eight men. The Ex. Ex. was the oceanic equivalent of the Lewis and Clark Expedition, a voyage into the last great terrestrial frontier, and it was a major force driving the growth of science in the United States.

In some ways, the Ex. Ex. could not have been more different from the POBSP. The POBSP was above all orderly and peaceful, its ships led by calm and reasonable naval officers. The Ex. Ex., by contrast, was commanded by an officer of the Captain Bligh variety: Lieutenant Charles Wilkes, a man who was seen by his sailors as hot headed and impulsive, moody and emotional, profane, violent, incompetent, and mentally unstable, who repeatedly ignored the advice of his officers. After his return, Wilkes was court-martialed for illegally flogging crew members, massacring the inhabitants of a small island in Fiji, and lying about when he first sighted the coast of Antarctica.

Unlike the POBSP survey, the Ex. Ex. was a true voyage of discovery: it definitively established that Antarctica is a continent. Previous explorers had sighted Antarctica but did not know whether it was a continent as opposed to an island, ice shelf, or something else. Wilkes proved that it was a true continent by sailing along, and mapping, a 1,500 mile stretch of the Antarctic coastline, a region now known as "Wilkes Land."

The projects also differed in scale. The POBSP crews visited forty-eight Pacific islands, all of them previously well charted. The Ex. Ex. was a tour de force of mapping and exact positioning. Its ships visited, surveyed, and mapped no less than 280 islands, many of whose geographical coordinates were only vaguely known beforehand.

Further, the Ex. Ex. was, if anything, overly dangerous, violent, and deadly. While surveying the Antarctic coast, two of the ships were often icebound, laden with ice and snow, lost in blinding gales, and blown backward into icebergs, while crew members were subjected to subfreezing temperatures for which they were not properly equipped.

But even in the face of those dangers, the Ex. Ex. made ample time for the peaceful pursuit of science. The personnel on board the ships included the naturalists Charles Pickering and Titian Ramsay Peale. Peale was the son of Charles Willson Peale, the portraitist, inventor, and founder of the Philadelphia Museum, one of the first museums in the United States. Also along was a geologist, James Dwight Dana, and a botanist, William Dunlop Brackenridge.

After an island was surveyed and charted, the expedition's scientists went ashore and explored it for artifacts of all kinds, which they collected aboard the ships in vast quantities. One of the returning vessels listed the following items as cargo: "Twenty five boxes and ten barrels of shells; twenty five boxes and one barrel containing botanical and other specimens; Seven boxes containing curiosities from Fiji islands; One box containing seeds and roots and eight boxes containing coral; One box containing Deep Sea water. One Fiji Drum; Thirty-six bundles and one Box containing spears and clubs; One box containing wheat. One box containing flower seeds; One box containing log books, one box containing a sleigh; One box containing books for philological department."

Added to this were anthropological and ethnographic specimens, items such as human skulls, mummies, weapons, body ornaments, baskets, grass skirts, and a feather blanket. In all, the Ex. Ex. crews amassed a hoard of objects estimated to have weighed a total of forty tons—approximately the weight of an adult gray whale. In fact, according to Smithsonian historian Jane Walsh, "the amount of material arriving at American ports began to reach overflow capacity." The final tally of the Ex. Ex. collections encompassed more than sixty thousand plant and bird specimens alone.

And so the US Congress, which had authorized the trip to begin with, now faced the question of what to do with the mass of artifacts and specimens that made it back to port. Congress would not officially establish the Smithsonian Institution until 1846, four years after the return of Ex. Ex. And the Institution's first museum, the "Castle," would not come into existence until later still, in 1855. Where to put the collection in the meantime?

Former secretary of war Joel Roberts Poinsett (after whom the poinsettia plant was named) proposed a novel solution. In 1840, before the end of the expedition, Poinsett had founded an organization called the National Institution for the Promotion of Science. He offered to have this new society take charge of the Ex. Ex. collection and house it in a just-completed structure in Washington: the US Patent Office Building.

And so in 1842, after having been inspected, cleaned, and cataloged, a select portion of the zoological, botanical, mineralogical, and other specimens from the Pacific and elsewhere were set out for public view in the Patent Office. In 1855, the Smithsonian Castle was completed and formally named the United States Museum of Natural History. And two years after that, Congress authorized the transfer of the Wilkes Collection from the Patent Office Building to the Smithsonian Institution's new exhibit hall.

Even today, the artifacts from the Ex. Ex. occupy a special place in the museum's holdings. "The Exploring Expedition collection," says Jane Walsh, "represents the earliest collected material in nearly every department in the Museum of Natural History, including botany, geology, vertebrates, and anthropology." Many of the original items are still on public view, more than 150 years after they were hauled back from the South Pacific.

* * *

The second prequel to the Pacific Ocean Biological Survey Program, the Tanager Expedition of 1923, was named for the decommissioned minesweeper, the USS *Tanager*, that carried the explorers from island to island. Their leader, Alexander Wetmore, would come to be a living link between the 1923 Tanager Expedition and the Pacific Program's biological survey of the 1960s. Wetmore was still alive and living in the Washington, DC, area during the Pacific Program and made himself, his field notes, photographs, and other unpublished materials available to POBSP field team members. Later, after Charles Ely, Roger Clapp, and Binion Amerson had written drafts of their respective accounts of the islands that Wetmore had visited, Wetmore read and commented on the history sections of those reports.

Widely regarded by birdmen as "the dean of American ornithologists," Wetmore was arguably the most bird-crazy secretary in the Smithsonian's history. He made his first written field journal entry at the age of eight, when on a trip to Florida with his family he recorded his observations of a pelican: "a great big bird that eats fish." His first publication, bylined "Alick Wetmore" when he was just fourteen, appeared in the journal *Bird Lore* in 1900, entitled "My Experience with a Red-headed Woodpecker." In it, he noted that he "was rather surprised to see that [the woodpecker] could easily go down a tree backwards, lifting his tail, and, after hopping down, falling back on to it."

The Tanager Expedition was a joint undertaking of the US Biological Survey, where Wetmore was employed as a field biologist, together with the Bernice P. Bishop Museum of Honolulu, and the US Navy. Its primary objective was to perform a natural history and biological survey of the Northwest Hawaiian Islands from Nihoa and Necker in the south to Midway and Kure at the northern end.

Several surprises were in store for Wetmore and his fellow researchers during the course of their travels. They discovered one new bird species. They also beheld, with their very own eyes, in real time, the extinction of another. The team members made some puzzling anthropological discoveries: for example, they found evidence of previous human habitation on Nihoa Island, in the Hawaiian chain, where there was neither a reliable source of fresh water nor sufficient living space for even a small community.

What most impressed the members of the Tanager Expedition, however, was the impact of human activity on the biology of many of the islands they visited. This would also be one of the chief lessons of the POBSP. Left to themselves, islands thrive, their native plants and animals reproducing, proliferating, and diversifying toward the limit of the island's natural carrying capacity. Frequently, however, the arrival of humans literally remade the face of an island, usually for the worse, although often as an unintended consequence of their actions. Obviously, people built structures and engaged in activities that displaced an island's life forms. And they introduced invasive species such as rats, cats, or rabbits that drove native species of plants, birds, and mammals out—or sometimes to extinction—and often to the point that the island stood in need of ecological restoration. In the Northwest Hawaiian chain, there is no better example than Laysan Island.

Laysan is a place with a history; it is in its way even infamous for being the Hawaiian island most thoroughly changed by successive waves of human occupation, despoliation, and destruction. The island is about one and a half miles long by a mile across, with a central lagoon. But the place's major feature was its birdlife; in 1857, an explorer described the original and pristine land area as " 'literally covered' with birds . . . [T]he birds were so tame that it was difficult to walk about the island without stepping upon them." In 1903 William Alanson Bryan of the Bishop Museum in Honolulu estimated that Laysan had a population of ten million seabirds, so many that they numbered "perhaps twice the inhabitants of Greater New York."

The millions of birds had over the years cumulatively deposited on the island's sands thousands of tons of guano (phosphatized carbonate of lime). Guano is high in nitrogen and therefore useful as a fertilizer and in making gunpowder—two of life's necessities at the time, apparently. Starting in 1890, a San Francisco mining company came to Laysan Island and turned the place into a guano excavation and transport facility. Ships chartered by the North Pacific Phosphate and Fertilizer Company brought Hawaiian and Japanese contract laborers to the site. They arrived with pickaxes, crowbars, and shovels to unearth the guano, and wheelbarrows to carry it away. Other workers built ramps for the wheelbarrows and, against all expectation as to what rightfully ought to exist in a tropical paradise, they laid down ties and tracks for a tramcar to haul the guano from the digging site to the shore. Its wooden cars were pulled along by a pair of mules. The workers also built houses to live in, a lighthouse, storage sheds to protect the piles of mined guano, plus a wharf from which to load the cargo onto the ships. For a period of a little over twenty years, until the early 1920s, this was a major Laysan Island industry, one that yielded more than 450,000 tons of guano before the boom ended.

As counterpoint to all of this debilitating work, the mining company's manager, Max Schlemmer, known as the "King of Laysan," also brought in an assortment of domestic rabbits, European hares, and guinea pigs to the island. This was largely for the entertainment of his wife and their many children, who lived on Laysan, but also as a source of fresh meat. Predictably, the animals reproduced exponentially and progressively ravaged the island's formerly abundant plant life, almost to the point of starvation.

Further ecological damage was done by Japanese millinery feather hunters. Feather hunting was a particularly heartless practice, as it normally took place during breeding season when sitting hens were confined to their nests. In 1909, feather hunters removed a total of one ton of bird feathers and two tons of wings from an estimated sixty-four thousand birds.

In response to these mass killings, President Theodore Roosevelt that same year proclaimed the Northwest Hawaiian Islands, including Laysan, as the Hawaiian Islands Bird Reservation, and placed it under the jurisdiction of the Biological Survey Division of the US Department of Agriculture. It was henceforth illegal "for any person to hunt, trap, capture, willfully disturb, or kill any bird of any kind whatever, or take the eggs of such birds within the limits of this reservation."

In the winter of 1912–1913, the US Biological Survey sent out a crew aboard the US revenue cutter *Thetis* in an attempt to exterminate Laysan Island of its rabbits. The shooters ran out of ammunition after killing five thousand rabbits, leaving a substantial number still alive.

When the Tanager Expedition set out from Honolulu on April 4, 1923, its first stop would be Laysan Island.

* * *

In addition to Wetmore, the expedition included several specialists— biologists, botanists, a herpetologist, and an archeologist. A professional photographer, Donald Dickey, was also aboard. To eliminate the rabbits from Laysan, which was a primary objective of the trip, the ship's complement included a pest control expert, Charles E. Reno. There were thought to be several hundred rabbits still on Laysan, and Reno proposed that the ship be equipped with at least two .22 caliber rifles, plus twelve thousand rounds of ammunition and seventy-five ounces of strychnine. In the end, the *Tanager* carried four rifles, fifteen thousand rounds, and an unknown quantity of the poison.

Wetmore was not looking forward to the arrival on what he thought would be a wasteland devoid of plant, bird, or mammal. After a three-day cruise, the island came within view on Saturday, April 7. Wetmore kept detailed, handwritten field notes of significant events and observations that he and others made over the course of the journey.

"It was with mingled feelings that I swept Laysan with my glasses," Wetmore wrote in his field notes. "As we lay a half mile away it presented merely a barren sand island rising a few feet above the water. Two coconut trees that rose in front of a half a dozen low tumbledown buildings with a low bush or two at either side were the only signs of vegetation."

Photographs taken by expedition members indeed show a desert isle (fig. 3.1). They resemble nothing so much as the pictures taken by the Apollo 11 astronauts, Neil Armstrong and Buzz Aldrin, of the moon's surface, which Aldrin described as "magnificent desolation."

Fig. 3.1 Laysan Island, April 1923
Credit: Bishop Museum Archives

Adding to the desolation was a row of wooden shacks abandoned by the guano miners: houses without windows or doors, their roofs caved in, their rooms filled with windblown sand. But amid all the emptiness and destruction, there were still plenty of birds.

"Birds were everywhere," Wetmore recorded. "Four or five Laysan Albatross stood on the sand of the porch with a fuzzy youngster half grown at the corner. Hawaiian Noddies rested on the roofs and window ledges, flying out with great clatter, while the White Terns rested on scant nests of *Sesuvium* stems, probably made by other birds, or perched about on the rafters or window sills inside. Wedge-tailed Shearwaters were everywhere, alone or in couples."

The photographer Donald Dickey kept a journal of his own, and in it he records that on April 11, 1923, Charles Reno sighted three Laysan Honeyeaters alive and well, the last known members of the species. Four days later, Dickey sighted two of the Honeyeaters scrambling through rocks in and among the leaves of a tobacco patch, hunting for insects.

On April 18, Dickey, now with his movie camera along, heard the Honeyeater again, "a weak but charming song behind me and [I] whirled to find one of our pair of Laysan Honey Eaters singing his heart out for me. Whirled the camera, slammed the focus lever, cranked and think I have him."

He did. The film exists today as a silent, ten-second-long YouTube video clip under the title "Laysan honeyeater, 1923."

On April 23, a storm came up. "Rain and mist—driving N gale—clearing to sandstorm in afternoon. Hell of a day."

The storm continued for four days straight. "Hell on earth," Dickey wrote in his journal during the worst of it. "Life in the open or in the tents or tumble-clown shacks is equally unbearable."

Finally the wind abated, but the island was a changed place afterward. Although many birds had survived unharmed, others had been blown off the island, and some two hundred dead carcasses were strewn about, covered by sand. The Laysan Honeyeater was never seen again.

The Tanager Expedition had both scientific and ecological payoffs. First, it provided species accounts of the birds populating the Northwest Hawaiian Islands. Second, it did a modest amount of restoration work on Laysan Island, planting seeds in an attempt to re-establish its once profuse plant life. Most important, Wetmore and his men succeeded in finally exterminating Laysan Island's rabbits. A visitor to Laysan in 1924 saw none. Neither did Binion Amerson, Fred Sibley, or any of the other field team members when they arrived on the island forty years later, in 1963.

* * *

By the time that first group of Smithsonian seafarers set out in February 1963, the army had identified its area of primary interest in the Pacific. It was a region that spanned both sides of the equator, extending from 30 degrees north latitude, just above the Hawaiian Island chain, to 10 degrees south latitude, about six hundred miles south of the equator. The area's western border was set at 180 degrees west longitude, which corresponds to the International Date Line, and the eastern boundary at 150 degrees east. The result was a great rectangle encompassing a total area of approximately 4.33 million square miles of mostly trackless, open ocean.

That wide expanse of sea embraces three main island groups, as well as two outliers that are not part of any grouping. The northernmost group consists of the Northwest Hawaiian chain. Dropping south, below the equator, are the Phoenix Islands, including Gardner, Sydney, Enderbury, and Hull. And the third is an array of islands that cross the equator: the Line Islands, so called because they stretch out in a line from Kingman Reef and Palmyra at the top to Jarvis, Malden, and Starbuck at the bottom.

The first of the two outliers *within* the great rectangle is Johnston Atoll, well south of the Hawaiians. Johnston Atoll had its own smaller rectangular grid area adjacent to it. This was a stretch of open ocean lying some 150 miles to the southwest of the atoll, called, variously, the Northern Grid, the fixed Johnston Atoll Grid, and Smithsonian Grid No. 1. The second was the twinned pair of Howland and Baker Islands, both lying a scant few miles above the equator. These two islands had their own separate grid surrounding them, and despite the fact that they were (just barely) in the Northern Pacific, it was called the Southern Grid, or Smithsonian Grid No. 2.

And then, finally, there was an island to be visited that lay completely outside the great rectangle: Eniwetok Atoll, in the Northern Pacific, far to the west of the International Date Line.

It was an oddity of the army's support of the POBSP that although the project's scientists visited and reported on all of these islands, as well as many others, the only ones on or near which the army conducted biological weapons trials were the outliers. The first trial was performed within the fixed Johnston Atoll Grid; the second on Baker Island, inside the great rectangle (and also inside its own grid); and the third on Eniwetok Atoll, *outside* the great rectangle. This was an index of just how far away from everything the Army wanted to get.

4

Life in the Field

For a project conceived and financed by an agency of the federal government and conducted by a large research institution, the Smithsonian's Pacific
Program got under way with an almost unseemly dispatch. The original contract between the Smithsonian and the army was signed in October 1962,
and the very first Pacific Ocean Biological Survey Program (POBSP) cruise
set out a little more than three months later, in February 1963.

Over the seven-year duration of the project, the fieldworkers would make
over a hundred separate voyages, going to almost every island they visited
not just once but repeatedly and in no easily discernible order. Further, the
names of the cruises bore no necessary relationship to where they actually
went. The Southern Island Cruises (SICs) were so called not because they
went into the South Pacific (although some did) but rather because they
sooner or later ended up traveling south from Honolulu, the starting point,
to islands that were located inside the Southern Grid. This was an expanse of
ocean that lay within a one-hundred-mile radius of equatorial Howland and
Baker. Still, many Southern Island Cruises began by first traveling north, up
through the Hawaiian chain, for weeks at a time.

Additionally, some of the Southern Island Cruises were also called "ATF
trips" since the field teams on those cruises traveled aboard tugboats of the
US Navy Auxiliary Tugboat Fleet, a flotilla of oceangoing tugs that could keep
up with a fleet of bigger ships during operations. In the Pacific Program, the
tugboats USS *Moctobi* (ATF-105) and the USS *Tawakoni* (ATF-114) would
carry Smithsonian scientists from island to island.

The very first of these cruises departed Honolulu on February 4, 1963,
and by any standard was an eventful, even unusual, voyage, perhaps because
it was the first of its kind—a prototype, or shakedown, cruise. Fred Sibley,
for one, thought it was "rather chaotic." It started with a month-long visit
to several islands of the Leeward chain, northwest of Hawaii, specifically
to Laysan, Midway, Kure, and Lisianski Islands, and to Pearl and Hermes
Reef. Afterward, the group headed south for a weeklong stay on Howland
and Baker.

Most of these voyages are well documented day by day, and sometimes even hour by hour, since many of the Smithsonian's biologists kept handwritten or typed field notes. Some were very detailed, Binion Amerson's journal particularly so, others less complete, but all of them described where the men went, what they saw and did once they got there, and occasionally even how they felt about it all.

A typical cruise consisted of five Smithsonian fieldworkers, as was the case with the first Southern Island Cruise, SIC 1 (ATF Trip No. 1), whose personnel consisted of Fred Sibley (ornithologist and trip leader), Binion Amerson (entomologist), Bill Wirtz (mammalogist), Allen Blagden (artist), and Bob McFarlane (ornithologist). For Amerson, Sibley, and Blagden, the trip started from the Smithsonian Institution itself. From there the three traveled together to BWI, Baltimore Washington International Airport, for flights to Honolulu via Chicago and San Francisco.

At Pearl Harbor the men boarded the ship that would be their home at sea for the next two months: the USS *Moctobi* (ATF-105). It was larger and heavier than an ordinary harbor tug; to the scientists, however, these ships were miserably small and confining. Fred Sibley, for one, dismissed the *Moctobi* as "the smallest seagoing ship in the Navy." This even though the *Moctobi* was 205 feet in length with a beam of 36 feet, with a crew of eighty-five officers and enlisted men. Its captain on this voyage was Lieutenant Commander Richard Owen Gooden.

At the Pearl Harbor shipyard, the men boarded the craft on January 28, 1963. Navy crew members showed them around their living, sleeping, and work quarters. The fieldworkers then went about stowing their personal supplies, clothes, and equipment. Binion Amerson set up and tested a Berlese funnel, a device for separating insects from soil samples and from bird nests. Meanwhile, shipyard workers installed a set of freezers for the bird sera, stomach contents, and the other biological samples that the fieldworkers were supposed to bring back with them or, alternatively, ship to the mainland.

As perhaps befit an expedition that was assembled in haste, its first few days did not go smoothly. Although the ship was scheduled to depart on February 1, the sailors discovered on that date that its anchor windlass was not operational and needed to be replaced. Mechanics brought a new one aboard and installed it.

"This did not function either," Amerson wrote in his journal. "Mid-afternoon we learned that we would not leave until the 2nd."

But on that day the trip was again postponed, until February 3.

On February 3, however, the ship still remained at the dock. "The ship-yard worked all night on the windlass and finished it by morning," Amerson wrote. "Due to it being a Sunday the skipper decided not to leave port until the 4th."

So the men rented a car and spent the day exploring the outer-edge sights of Oahu, visiting Diamond Head, Koko Head, and Sacred Falls, among other tourist attractions. They took photographs, sighted a few whales in the harbor, and went swimming at Hauula Beach. The water there was very clear, they thought, but the sea bottom was quite rocky.

On the following day, at last, February 4, the *Moctobi* did indeed steam out of port.

The fieldworkers were charged with collecting data at sea as well as on the islands, and so the men started recording their first bird observations right away. They spotted many examples of the pelagic birds common to the area: black-footed albatross, Laysan albatross, sooty terns, a frigate bird, and some jaegers—hawk-like seagulls.

"We shot and collected two of the jaegers," Amerson wrote. "In order to do this, the ship had to back up. The birds were picked up by a net on a long bamboo pole."

Next morning, the ship's main bearing burned out, meaning that the *Moctobi* would have to return to Pearl Harbor for more repairs, and the men for yet more Hawaiian sightseeing. Nevertheless, they continued to count and identify birds all the way back to port. Before noon, they had also captured a black-footed albatross in flight with a hook. They banded the big bird and released it.

At length, after all of these postponements, delays, and a false start, the ship sailed out of Pearl Harbor again on February 8. "This was the only time in nine trips the Navy didn't depart within an hour of the scheduled departure time," Fred Sibley remembered afterward.

Their first port of call would be Laysan Island, as had also been true of the Tanager Expedition. Laysan was about nine hundred miles northwest of Honolulu, and the voyage took three days. The men got seasick one after the other.

"Sibley began to get sick before lunch and Wirtz and Blagden got sick during lunch," Amerson wrote. "I did not get sick until after eating dinner, which did not agree with me."

"We had experienced 40–50-degree rolls on the trip," Sibley explained. After this, the SIC cruises were sometimes referred to as "SICk cruises."

The *Moctobi* arrived at Laysan on February 11, 1963. The men made an easy landing by rubber raft in relatively calm conditions, after which they established a campsite, pitched their tents, and then went off to survey the island. The terrain they now strode across was nothing like the desolate place that had greeted Alexander Wetmore and his team forty years earlier. Instead of a barren wasteland, Laysan, with its rabbits long gone, was now covered in vegetation, birds, seals, and sea turtles. It was more like the legendary "tropical island paradise" of popular imagination.

Over the course of the POBSP's several visits to Laysan, the scientists would identify forty-two species of plants on the island, some of which had been introduced by humans, either deliberately or accidentally, species such as onions, potato, tobacco, and Bermuda grass. In addition, there were coconut palms and casuarina trees, plus dense stands of bunchgrass, sedges, shrubs of various types, luxurious mats of sea purslane, flowering hibiscus, morning glories, nightshade, scaevola, and several other flowering plants. The place did not lack for variety or richness of plant life.

Animal life on the island had once been on the brink of extinction. As Roger Clapp and Charles Ely wrote in their POBSP report on Laysan:

> The combination of tolerable living conditions at the guano headquarters and the remarkable biota early attracted several biologists and their reports soon made Laysan synonymous with teeming colonies of fearless seabirds—a veritable paradise for biologists. Unfortunately island occupancy also resulted in a continuing conflict between the native biota and man and his introductions. In less than 35 years this conflict resulted in the destruction of two endemic plants (and nine other native species), three endemic birds, and a number of the endemic insects. Other populations, notably those of seals and turtles, were gravely depleted and only the timely arrival of the Tanager Expedition prevented complete destruction of the vegetation and the entire endemic biota.

But the island had since recovered completely. Over the POBSP's fourteen separate visits to Laysan, the field teams recorded the presence of nineteen breeding species on the island, many of them in great numbers. There were some two million sooty terns there alone. The Laysan albatross was the second–most abundant species on the island, at well over a half million. There were wedge-tailed shearwaters and Bonin petrels, each numbering in the hundreds of thousands. Black-footed albatross, Bulwer's petrel, and

Christmas shearwater each numbered in the tens of thousands, with lesser numbers of several other species also present and breeding. In addition, there were numerous examples of other species that visited the island on a transient and irregular basis.

The only mammal found there was the Hawaiian monk seal, with more than three hundred observed on the island's sands. Finally, there were also a few snake-eyed skinks hiding among the rocks. In all, this was an island once again vibrant with life.

* * *

The real work of the Smithsonian field teams, aside from counting birds, lay in banding as many specimens of as many species of birds as possible. The first were Bonin petrels.

"We soon learned they could bite like hell and could draw blood," Amerson wrote in his journal. But he also learned that he could immobilize their heads between his forefinger and middle finger to keep from getting bitten.

In principle, banding was a straightforward process. It involved sliding a numbered plastic or metal band up the right leg of a bird, and then recording the number, species, and any other information that was readily available, such as the bird's sex and whether it was an adult or chick. To keep it simple, birds of the same species, sex, and degree of development were banded in lots of one hundred.

In practice, there were a few tricks involved. Most of the POBSP banding work was done at night, and the bander wore a small headlamp that temporarily blinded the birds and made them easy to catch. The partially opened bands came in groups of one hundred on long plastic tubes. First, the bander grabbed the bird with his left hand, holding its bill away from him and its right leg stationary. He removed a band from the plastic tube and slipped it onto the leg. Then, with a pair of special pliers, he closed the band securely around the leg, double-checked its fit, and immediately released the bird. Right away he reached for another one with his left hand and repeated the process time and again. Often, the birds were so tame and so densely clustered that the bander didn't have to move far, if at all, to obtain the next in sequence.

In 1965, an anonymous member of the POBSP (but almost certainly Roger Clapp), wrote an account of his banding work for the group's monthly newsletter, the *Pacific Bird Observer*, and, after describing the process commented that, "In this manner [a skilled bander] can band 500 birds in an hour under ideal conditions—approximately one bird every seven seconds. After several hours of such work, he will have blisters on his hands from using the pliers, his old cuts will have been opened by the birds' pecking and scratching, he will be covered with a light coat of regurgitated fish (and other by-products), and he will wonder why he ever decided to work with birds."

Bird banding was anything but fun, glamorous, or exciting. It was repetitive, grueling, exhausting work. And, for it to have any point at all, at least a few of the birds would have to be recaptured elsewhere and the band number reported to the proper agency.

"Much of the success of the Pacific banding program depends on the many observers in the Pacific Basin who report banded birds," the anonymous author wrote. "We consider ourselves fortunate if one Sooty Tern out of a thousand banded is recaptured away from its banding site. Even with larger birds such as the Lesser Frigatebird it is unusual to have recoveries of more than one out of a hundred."

Binion Amerson was a newcomer to the banding process but learned to do it quickly enough, and within a couple of hours of starting the work on Laysan, he and the rest of the men had banded some three hundred Bonin petrels. They found that doing the work at night was a mixed blessing. The darkness made it easier to approach and catch the birds, but it also obscured the sand burrows the petrels dug and used for nesting.

"As we walked about the island we constantly fell into the petrel burrows," he said. "At times we were up to our knees in the sand. This was quite annoying."

The field trip members stayed at Laysan for a couple of days, and then they were back again at sea. Because of the prevailing winds and waves, the men had landed on one side of the island but were forced to depart from the other, which meant that they had to haul all of their gear, tents, and other equipment across the width of the island, which was a distance of slightly more than a mile.

The next stop was Midway Atoll, where the men faced a task that went beyond banding: they were to collect 150 Laysan albatrosses, put them in boxes, and send them by plane more than five thousand miles to Fort Detrick. This would be possible because Midway was already the site of a military airbase. Indeed, Midway was the best known of the Northwest Hawaiian Islands, made famous by the Battle of Midway in June 1942, which involved hundreds of aircraft, several aircraft carriers, two battleships, and more than eighty cruisers, destroyers, and submarines. This engagement, six months after the Pearl Harbor attack, proved to be a decisive American victory over the Japanese, and it effectively turned the tide of World War II in the Pacific.

Midway was so named because it was roughly equidistant from North America and Asia. In 1859, Captain N. C. Brooks of the sealing ship *Gambia* claimed the island as a possession of the United States under the Guano Act of 1856. For complex reasons, Midway was not made a part of Hawaii but remained under the jurisdiction of the federal government.

Nor was Midway included in Theodore Roosevelt's executive order as one of the islands that made up Hawaiian Islands Bird Reservation. Instead, it became both a commercial outpost and a military facility, due in large part to its strategic location in the Pacific.

The atoll consists of three islands: Sand Island (1,200 acres), the much smaller Eastern Island (334 acres), and the tiny, no-account Spit Island, whose acreage is so small that it varies with the tides. In the early years of twentieth century, the Commercial Pacific Cable Company set up operations on Sand Island as part of an attempt to lay an international telephone cable across the Pacific. Between 1906 and 1920 the company brought in beach grass to control erosion and ironwood trees to provide windbreaks. For a decade, the company also imported shiploads of topsoil. Pan American airlines, in support of its transpacific service, imported yet more topsoil, from Guam. In addition, Pan Am built an airport, terminal building, and even a hotel on the island. Later came roads, streetlights, even a movie theater.

Midway also hosted the Pacific's largest colony of Laysan albatrosses, almost a million of them. Because it also harbored a naval base, when the POBSP men landed at Midway they had no need to pitch tents but rather had the luxury of staying either in airbase housing or remaining in their quarters aboard the *Moctobi*.

The POBSP field team had several objectives at Midway, apart from counting, banding, and collecting live birds for shipment to the mainland. In addition, they were to gather live ticks to be shipped to a Dr. Guilford, at Detrick. The latter task fell to Binion Amerson, the group's insect expert. Finally, he and Fred Sibley would kill, skin, and pack into an empty trunk about a dozen gulls, together with an unknown number of skeletons that were to be sent back to the Smithsonian as specimens.

Collecting the Laysan albatrosses was easy enough, given the island's large population of them. Preparing them for shipment and getting them safely to the States was another matter, as the birds were not small. The stuff of legend, great albatrosses have the longest wingspans of any bird, up to twelve feet from tip to tip (fig. 4.1). Although they nest on land, they spend much of their time in soaring flight or floating on the sea, and while on the water they have been known to lose their legs to sharks.

On March 9, 1963, the Deseret Test Center's John Bushman arrived at the Midway air base to help capture and pack the albatrosses for shipment. Over the next two days, Bushman, Amerson, and Sibley collected some 150 Laysan albatrosses and started boxing them. (The previous day, Allen Blagden, the artist, left on a return flight for the States to attend to his father, who had been hospitalized.)

The boxes came in by ship but had gotten wet in the forward hold and were in bad shape, practically falling apart. To repair the damage, the men put a piece of dry, waxed cardboard on the bottom of each box. Then they packed the birds into them, at which point the albatrosses were ready to go.

Fred Sibley, who would accompany the birds all the way to Washington, describes the further sequence of events. "When loading the albatrosses, some twelve hours after being boxed, we formed a line and passed boxes from the truck to the plane. The albatrosses, having powerful bills, had

Fig. 4.1 Binion Amerson holding a Laysan albatross, Midway Atoll, 1963
Credit: Jeff Cox

chewed through their cardboard boxes so they could poke their heads out. At one point I reached back for a box and put my hand in the beak of the next albatross. In the course of learning to band albatrosses at Midway I had acquired a multitude of minor bites. However, this time I made the mistake of immediately pulling my hand away when bitten and ended with a huge gash between thumb and forefinger."

At length, however, the birds were ready to go, on what was supposed to be a series of secret flights to Honolulu, the West Coast, and finally Washington. A small contract airline took the birds from Midway to Honolulu, arriving at the military side of the airport. "These bird shipments back to Washington were 'secret,'" Sibley recalled. "But ignoring the smell of 150 birds, who have spent twelve hours in boxes is impossible on a smallish plane with passenger compartment open to the cargo hold. Thus a lot of people ended up knowing there were birds on board."

The military then brought the birds to the Pan Am terminal for transport to San Francisco or Los Angeles, and from there put them on another flight to their final destination. This meant two transfers.

"At each transfer you had to go on the tarmac and check that birds were being transferred and not left sitting in the sun somewhere," Sibley remembered. "All the people were very helpful and bent all sorts of rules—after all who wants 150 dead albatrosses in their warehouse?"

How many of the birds actually wound up, dead or alive, at Fort Detrick was unknown to the field team members. As was the purpose of sending them there to begin with.

"No idea what happened to the birds," Fred Sibley said. "Testing reaction, resistance, etc. to various biological agents?"

The remaining members of the POBSP's first ATF Cruise made additional stops at the rest of the Northwest Hawaiian Islands, including Lisianski and Pearl and Hermes Reef, before steaming south to Howland and Baker. By March 31, the men were back in Washington for their two-month rest period. All of them would make further POBSP trips into the Pacific except for Allen Blagden, who by this time had quit the program.

* * *

By the time the ATF Trip No. 1 ended on March 30, 1963, the Smithsonian researchers had banded a total of 8,179 birds. Between 1963 and 1968, POBSP biologists would make thirteen additional expeditions to Laysan Island, in some cases visiting three times per year. Amerson himself came back only once. But for all of these visits, bird counts, and bandings, the US Army took no interest in Laysan Island and conducted no tests there, biological, chemical, atomic, or otherwise. Which in the end would be true of most of the islands, atolls, reefs, pinnacles, and other bits of land explored by members of the POBSP.

Of the multiple objectives of the Pacific Project, one was to collect blood samples from birds found on the islands as well as from those caught at sea. These specimens were sought after by Fort Detrick scientists because they were interested in finding out what kinds of viruses, bacteria, or antibodies, if any, might be harbored by different species of birds.

In August 1962, a couple of months before the Smithsonian–Detrick contract was signed, Peter J. Gerone, of the Virology and Rickettsia Division at Fort Detrick, produced a memorandum, "Collection of Bird Sera," that was later used as a guideline by the POBSP fieldworkers. The memorandum described the process in detail:

> The following procedure should be adhered to by a prospective contractor for the collection of bird sera to be used in HI [hemagglutination-inhibition] tests with Group A Arbor viruses:
>   a. Birds should be captured without trauma. [As if!]
>   b. In the case of species in which individual birds will yield 2–3 ml of whole blood, individual serum samples should be taken. If there are smaller birds of other species which will not yield 2–3 ml of whole blood, then bloods from not more than 3 individual birds should be pooled.

The memorandum went on to give instructions for the centrifugation and refrigeration of the blood, placement of the vials into sealed containers, and precise labeling information. It then continued:

>   e. Serum samples should be stored in a household deep-freeze (−10 to −20°C) and shipped to Fort Detrick in the frozen state.
>   f. An attempt should be made to obtain 200 or more serum samples (individual or pooled as the case may be) as equally divided among the bird species in question as is practicable.

Given that blood sera obtained in this manner could be used to identify the viruses harbored by bird species, the question arises why Detrick scientists nevertheless wanted live bird specimens to be brought back to their laboratories. One answer is that they needed live specimens to determine whether a given virus or bacterium could be transferred from one individual bird to another member of the same species that had not already been infected by the pathogen. Indeed, an experiment of this sort was conducted at Dugway Proving Ground with wedge-tailed shearwaters and black-footed albatrosses, in 1964.

Yet another goal of the Pacific Project was to obtain bird skins of several species. This was desired not so much, if at all, by the military but rather by the Smithsonian for its own collection, which then numbered more than 625,000 individual specimens. Bird skins are valuable not merely for mounting and display but also for research purposes. Skins that are representative of a species are often needed for the reliable identification of a new-found bird, and in the case of Pacific seabirds, specimens are useful in the study of inter-island variations among members of the same species.

There is no getting around the fact that in the majority of cases, obtaining a bird skin requires killing the bird, although it is also possible to start with a specimen that is recently dead of natural causes. The most "humane" way of killing a bird is by manually compressing the lungs, a method that has the further advantage that no other damage is done to the body or feathers, as would normally be the case if it were shot.

In essence, the process of turning a dead bird into a bird skin involves eviscerating the specimen of its internal organs, removing any fat adhering to the inside of the skin, and filling the skin with a material such as cotton or excelsior to retain the original size and shape of the body. It was of primary importance in all this not to disturb, discolor, or otherwise soil the feathers.

"It's not that gruesome," Roger Clapp said of the overall procedure. Clapp was a master of the bird skinning art, and members of the POBSP would sometimes hold "skinning bees" in the field, in which an expert like him could process three small birds in an hour, start to finish.

Clapp's first ATF trip was the third POBSP Southern Island Cruise, which took place between October 3 and November 26, 1963. The Smithsonian crew in this case was Fred Sibley, trip leader; Roger Clapp, assistant leader; plus research curators David Bratley and Larry Huber, the latter of whom was also making his first foray into the Pacific. By this time, the trips were followed a more or less established pattern, although there was always the possibility of surprises or other random departures from the normal.

The ship in this case was the USS *Tawakoni* (ATF-114). It was the same kind of oceangoing tug as the *Moctobi* and was the same 205 feet in length, with a crew of eight officers and sixty-eight enlisted men. The Smithsonian fieldworkers found their quarters cramped and confining. "Packed in like sardines," Clapp recalled. "Sleeping below decks, oil smell all over."

"Had a lot of 'firsts' today," Huber wrote in his field notes on the trip. "First time I was on a Navy tug, first time I've eaten on a Navy ship, first time I've gotten sea-sick. I've also got my 'firsts' in sea birds: White-tailed Tropicbirds, Wedge-tailed Shearwater, Bonin Island Petrel, Brown Booby and a Red-footed Booby."

Seven days out of Hawaii they neared their first landfall, which was to be on the storied and mythic Howland Island. This was the island on which Amelia Earhart was to have landed for a refueling stop on her 1937 round-the-world flight attempt. In the end, Earhart and her navigator, Fred Noonan, missed the island completely. Still, a monument in the shape of a lighthouse was erected there in her memory—the so-called Earhart Light, also known as the day beacon. It had recently been repainted and fitted with an automatic light.

At 8:58 on the morning of October 9, the island itself came into view. It was an exciting and mesmerizing sight after many long and uneventful days at sea. The island was long, flat, and almost featureless. It was so thin and smooth, with an elevation of about ten feet, that from a distance it did not really look like an island, more like a thin, straight line floating on the water. In the middle of the expanse was a tiny, cylindrical projection: the day beacon.

"A big amazing thrill, here's this island out in the middle of absolutely nowhere," Clapp remembered much later. "We knew it was the Amelia Earhart island. That was part of the charm. Covered with birds. You can walk up to them. When you walk ashore some of them actually sit on your shoulder. No people live there."

Landing was an easy matter. "Beached on the island at about 10:30 and set up camp," Clapp wrote in his journal.

"Camp" on these islands was a decidedly primitive affair. Its centerpiece was merely a tent, and by no means a fancy one at that, especially during a tropical downpour.

"The only waterproof part of our tent turned out to be the floor," Fred Sibley remembered of his Howland Island shelter. "When water got four inches deep inside, we cut the floor."

The men hauled a few large stainless-steel canisters of drinking water onto the beach and then to the tent, along with their food supplies.

"Onshore we largely ate C-Rations, a diet that led to digestive problems, probably because most had passed their expiration dates, sometimes by nearly a decade," Roger Clapp said.

The men sometimes erected a shade canopy above a makeshift table. During meals the workers sat on portable chairs or wooden benches, and when the sun was beating down on an equatorial island such as Howland, Baker, or Jarvis, the men would often be dressed in nothing more than shorts and a T-shirt. The diners would be surrounded by, variously, clumps of sea grass or other flora, sand, driftwood, rocks, remnants of materials leftover from previous island residents or from plane crashes, and a scattering of birds. Occasionally the men would hang clothes out to dry on the canopy's guy lines.

Overall, the island living quarters of Pacific Project personnel resembled nothing so much as a holiday beach party. Not that it felt like a holiday to those living in such conditions for days at a time.

* * *

When Clapp and the others reached Howland in October 1963, not many artifacts remained of the island's long and full history. Captain Daniel McKenzie of the New Bedford whaler *Minerva* in 1827 or 1828 gave the island its present name, after the owners of his ship, I. Howland Jr. and Company.

After the island became known among seafarers, several stopped there to fish, as well as to gather birds and eggs for food. Later still, the island was again "discovered" by guano mining companies, several of which laid claims to, and started mining, its guano deposits. According to an unpublished report on Howland written by Fred Sibley and Roger Clapp in 1965, "It seems evident from reading early newspaper accounts that guano loading operations on Howland Island were quite perilous. In 1864 alone, three ships, the Mary Robinson, Arno, and Monsoon were wrecked on the island, and another, the White Swallow, slipped from her mooring and disappeared without a trace."

The rise of long-distance commercial airline flight in the 1930s, and particularly the advent of Pan Am Clippers that could land on the water, produced a renewal of interest in the equatorial "HBJ" islands, Howland, Baker, and Jarvis. All at once, these isolated specks of land became strategic assets. The US government decided that its claims to ownership would be best demonstrated by physical occupancy, that is, by placing American citizens on them as colonists. This would also serve the additional function of keeping Japanese forces away from the islands. For these reasons, in 1935 the US Department of Commerce enacted the American Equatorial Islands Colonization Project. Soon thereafter, three small groups of young Native Hawaiian men, many of them recent graduates of the Kamehameha School for Boys in Honolulu, took up residence, one group apiece on Howland, Baker, and Jarvis.

The Howland Island colonists had at their disposal a couple of prefabricated wooden structures built by the US government. The two houses were set behind a protective stone wall, and the settlement was given the name Itascatown, after the ship that brought them there: the coast guard cutter *Itasca*.

This was the same ship that, two years later, would be in radio contact with Amelia Earhart as she tried to catch sight of the tiny island from the air, which was difficult because scattered clouds left shadows on the ocean's surface that were visually indistinguishable from the island itself. And it was to the *Itasca*, standing offshore Howland, that Earhart transmitted what is generally accepted to have been one of her last messages, saying: "KHAQQ [her aircraft's call letters] calling *Itasca*. We must be on you but cannot see you." Shortly after that she became the most completely disappeared famous person in all of history.

When they arrived on Howland in 1963, none of the men, Sibley, Clapp, Bratley, or Huber, spent any significant time exploring the ruins of Itascatown, which in any case had been reduced to lines of rubble left over from the original stone wall. The remnants of the three airstrips that had been plowed and graded for Earhart were still clearly visible, however: long, wide stretches of sand relatively bare of vegetation.

They soon began work. "Started collecting birds for blood," Clapp wrote. "Got samples from ten masked boobies, four red-footed boobies, one *F. minor* [*Fregata minor*, great frigate bird], and three red-tailed Tropicbirds. All survived the heart puncture, but two masked boobies succumbed, one while I was holding it; perhaps from shock." (So much for collecting blood samples "without trauma.")

Huber meanwhile picked up lizards. "Caught about thirty skinks," he wrote. "These skinks out here are either a species with a subspecies or come in two color phases. Caught one gecko."

The work of the four men soon devolved into matters of routine: counting birds, eggs, and nests; banding birds; skinning them; taking blood samples; collecting insects. This did not mean that nothing interesting happened.

One night, Clapp said, "we received an urgent warning that a tidal wave or series of tidal waves plus an earthquake was imminent. Went to the lighthouse [the day beacon] and spent an uncomfortable hour or so clinging to the top. . . . Nothing happened but we had to remain at the base of the lighthouse until the all clear at about 12, which pretty well shot the night."

"We saw signals from the ship," Huber wrote. "They were using their blinkers and 12 inch spot light to attract us. They seem desperate. When we got back about 30 minutes from seeing the first message, we found out that a tidal wave from Japan was coming and that we should go at once to the lighthouse on the island (only 20 feet tall). Didn't notice any tidal wave."

Later on the same ATF Cruise, Clapp and the others landed on Birnie Island. Birnie is a member of the Phoenix Islands, a group lying just below the equator. Even for a "minor outlying island," Birnie is small, just forty-nine acres, smaller than the average eighty-acre American farm. Guano companies had once staked out claims to the place but never did any actual mining there

When they arrived on Birnie in the fall of 1963, the group was the first scientific team known to have set foot on shore. As it turned out, what they saw there was not all that impressive.

"Birnie did not have much on it," Roger Clapp said. "Boobies, blueberry noddies, very little vegetation. Not many bushes, and some of these species are obligate bush nesters. Black noddies nest in bushes or stone walls. They would come in from the sea at night and set up in these little packs. Quite a barren island, not much there. We counted and sampled everything in two days."

Also, Birnie was crawling with rats. "Rats are everywhere," Fred Sibley reported, "and are most active after dark although they can be seen at all times of the day."

Still, practically everything they saw on Birnie was new to science in the sense that the island had never before been systematically observed and explored by trained researchers. To be part of the first scientific team to behold such sights was therefore a kind of privilege. It was, in its way, like being one of the earliest explorers of the Pacific.

Altogether, SIC 3 was a success. The four men visited a total of ten islands: the eight comprising the Phoenix group, plus Howland and Baker. In all, they banded 16,585 birds, collected 188 additional birds as specimens, and took 477 blood samples. They had been away for two months, and, after they got back to Honolulu, would be off for two. For better or for worse, like it or not, this was what day-to-day life was like in the POBSP.

5

The Artificial Atoll

Most of the forty-eight separate islands, atolls, pinnacles, and reefs surveyed by the Pacific Program scientists were "natural" in the sense that they were created exclusively by natural forces, had existed across long stretches of time, and remained largely intact during the modern era. This does not mean that they were static, stable, fixed, or eternal objects: islands are by nature dynamic, mutable ecosystems.

To take an extreme case, East Island, part of the French Frigate Shoals in the Northwest Hawaiian chain, was visited by members of the Pacific Project at least nine times between 1963 and 1969. It was a small island, only eleven acres or so, and had also been visited by Alexander Wetmore during the 1923 Tanager Expedition.

"The island rises from 8 to 10 feet above sea level and supports seven species of plants," Wetmore wrote in his journal. "The island has been much larger but has been cut away by storms. . . . Shearwater, Blue-faced Boobies, noddies, and young albatross are all about and furnish an endless source of interest."

In 1944, the US Coast Guard built a LORAN (long-range navigation) station on the island, together with several support buildings including barracks, a mess hall, a generator hut, and a machine shop. In addition, workers had erected several hundred-foot-tall radio antennas there. In 1952, however, the coast guard abandoned the island.

When the Smithsonian scientists arrived on East Island in 1963, the buildings were in ruins, but the tall antennas still stood upright. "These provide excellent island markers when approaching by small boat," according to the Pacific Ocean Biological Survey Program (POBSP) report on the French Frigate Shoals. The scientists found several species of birds on the island, among them more than ten thousand wedge-tailed shearwaters and almost two hundred thousand sooty terns. Small numbers of green sea turtles and Hawaiian monk seals were also present.

Much later, in October 2018, Hurricane Walaka passed over the French Frigate Shoals. It wiped away East Island almost completely, together with all of its structures, vegetation, and bird populations. The only vestige of the place remaining was a 150-foot-long patch of sand.

Six miles away, Tern Island, about the same size that East had been, was hit by the same hurricane. It nevertheless remained largely unchanged in size and shape; this is because Tern Island is an artificial island, designed and built to withstand the most destructive forces of wind and wave.

When Wetmore visited Tern Island in 1923 he described it as "about 600 yards long by 150 yards wide. The eastern half is a long curving sand spit horn 6 to 8 feet above the sea which is swept in time of storm. The western half, which is the site of the bird colonies is from 10 to 12 feet above the sea and has a sort of fine coral sand."

In the beginning, Tern Island was all of eleven acres, and it was a breeding ground for native birds. It had in fact been named by Wetmore for its great profusion of sooty terns—approximately seven thousand at the time of his visit. The island also hosted sea turtles, monk seals, and about three acres of vegetation.

Later, after Pearl Harbor, the military wanted to establish an aircraft refueling base and emergency landing field between Honolulu and Midway Island, and the French Frigate Shoals was about halfway between the two. On June 12, 1942, at the request of the military, Edward Brier, chief engineer of the Hawaiian Dredging Company, made a secret reconnaissance trip to the French Frigate Shoals aboard a navy vessel to evaluate the prospects of turning one of its many islands into an airport. The ship anchored a few miles from Tern Island and sent Brier across on a small boat. Once ashore, Brier found the place "covered by a few blades of tough grass and populated by tens of thousands of terns."

Those terns were soon gone. A little more than a month after Brier's visit, Project ME-36, the navy's code name for the transformation of Tern Island into an airport, began. On August 13, 1942, a US naval construction battalion better known as the Seabees arrived on Tern Island, together with bulldozers, cranes, dredging machines, and other equipment. They stripped the island flat and added a total of 660,000 cubic yards of coral fill from the surrounding waters. This increased the length of the island from 1,800 feet to 3,100 feet, and its size from eleven to fifty-seven acres. The Seabees also reshaped the island from a gentle crescent into a rectangle with precisely squared corners.

Most important of all, to buttress its perimeter against storm surges and tidal waves, the workers drove some five thousand sheet steel pilings fifteen feet into the seafloor. The tops of the pilings stood six and a half feet above the mean tide level. The pilings were concentrated at the ends of the island and were arranged in such a way that each end was perfectly straight across its width.

When finished, the enlarged and reshaped island resembled nothing so much as an aircraft carrier awaiting a squadron of planes. Indeed, a 1947 article about the project in *Military Engineer* referred to the result as "a stationary aircraft carrier." The resemblance was reinforced by the fact that the island now had not only a long "flight deck" but also "storage decks" in the form of bulges along the sides where as many as twenty-two aircraft could be parked at a time (fig. 5.1). To further the illusion of a ship in motion, when a strong current was running, a trail of turbulent water at one end of the "coral carrier" looked very much like the wake of a ship moving through the sea.

By the time the POBSP arrived there in June 1963, sooty terns and other birds had long since returned en masse. In fact, they began to straggle in shortly after the start of operations in March 1943, with the arrival of a few Laysan albatross, shearwaters, and white terns.

Between June 1963 and June 1969, Pacific Project field teams would make a total of eleven visits to the French Frigate Shoals, ten of them stopping at

Fig. 5.1 Tern Island, the "Coral Carrier," French Frigate Shoals, 1945
Credit: Frances Frezza, Naval Air Station Fort Lauderdale

Tern Island. During this time field team members banded a total of 67,027 birds of nineteen species, including 37,895 sooty terns, and roughly five thousand each of brown noddies and wedge-tailed shearwaters, plus four thousand black-footed albatross. There were also 8,238 recaptures of banded birds. Of that total, 7,958 had been banded on the various other islands of the French Frigate Shoals, while only 280 had been banded elsewhere.

During all the years of construction activity on East and Tern Islands, and amid all the disruptions brought upon their birdlife, those two islands, and the French Frigate Shoals as a whole, remained parts of the Hawaiian Islands National Wildlife Refuge. A US Fish and Wildlife field station was active on Tern Island from 1979 to 2012, when all personnel were evacuated in advance of a storm. In July 2013, the Google Street View project imaged the now-abandoned "Coral Carrier." A complete 360-degree view of the island, with its thousands of birds on the runway and everywhere else, can be seen online (https://tinyurl.com/uxshbycc).

In the wake of the hurricane that swept away East Island, Tern Island remained largely unchanged in its contours, although its surface structures and wildlife were substantially altered. A damage assessment report by the National Oceanic and Atmospheric Administration stated:

> At Tern Island, storm surge deposited sand and debris across the island, swept away vegetation, caused erosion, and changed habitat conditions. Portions of the island were completely overwashed or inundated by the ocean. Plants such as beach heliotrope (*heliotropium foertherianum*) were uprooted, burrows of nesting seabirds were flooded, and infrastructure left behind from the island's days as a U.S. Navy airfield in World War II and a U.S. Coast Guard Long Range Navigation radio station were significantly damaged or destroyed. Unfortunately, some seabirds were also killed and turtle nests washed away by the storm.

Today, Tern Island is part of the memorably named Papahānaumokuākea Marine National Monument, is once again uninhabited and is ruled by the birds.

* * *

But Tern Island, as artificial as it was, proved to be only an overture to Johnston Atoll, a four-island archipelago that was almost entirely synthetic. Probably not coincidentally, the island grouping was also the most intensively studied of all the sites ever visited by the POBSP. Between 1963 and 1969, twenty-nine of its field team members would make at least forty separate visits to Johnston Atoll, many residing at one of its islands for long stretches of time. The project's scientists banded more birds at Johnston Atoll than at any other location in the Central Pacific. And to facilitate future sightings of birds banded at Johnston Atoll, the scientists tagged them with blaze-orange plastic leg streamers.

Several factors explained why Johnston Atoll came in for this special treatment. Primary among them was the location. Johnston, said geographer Mark Rauzon, is "one of the most isolated atolls in the word." Unlike islands that were in groups such as the Line and Phoenix Islands, or the successive islands that made up the Hawaiian chain, Johnston Atoll had no near neighbors. It was off by itself, alone, the only dry land in more than 450,000 square miles of otherwise open and empty ocean. It was the oceanic equivalent of being "lost in space." The atoll was 717 miles southwest of Honolulu, and the nearest landmass of any type was the French Frigate Shoals, about 450 miles to the north.

Its remoteness from the rest of the world thus provided the ultimate in seclusion, and if anyone was seeking a place to conduct an operation in secrecy, hidden away from prying eyes, Johnston Atoll would fill the bill to perfection. Indeed, the privacy it afforded made its islands a magnet for those who wanted to do hazardous testing or dangerous work of any type, or who planned to perform scientific experiments whose existence and nature were best kept out of sight. (In fact, for a long time the atoll effectively kept itself out of sight. Spanish explorers and traders sailed the waters of the Central Pacific for two and a half centuries before any sightings of Johnston Atoll were recorded.)

The atoll was a US possession, and several agencies of the government planned to conduct many different types of experiments at Johnston Atoll or in the open ocean waters nearby. To serve all these disparate purposes efficiently, the government remade and rebuilt the atoll's original land areas and created still other, wholly artificial islands where none had existed before.

Indeed, virtually none of what exists at Johnston Atoll today, or when the Smithsonian's Pacific survey was in progress, was there to begin with. The definitive POBSP report on the place gives a succinct description of its makeup as of the mid-1960s: "Johnston Atoll consists of two highly modified natural islands and two completely man-made islands totaling about a square mile in surface area lying on a 14 by 7 mile coral reef platform in the tropical Pacific Ocean at 16°45'N, 169°31'W."

The larger of the two "highly modified" islands was Johnston Island itself. Originally a shoe-shaped isle of forty-six acres, its surface area was first graded flat and then enlarged three consecutive times (in 1958, 1962, and 1964) with coral dredged up from the sea. In the end, it was a 596-acre landmass roughly in the shape of a rectangle. Its primary use was as an airport for military transport planes and, later, commercial airline flights. In 1969 and 1970, Continental Airlines ran an "island hopper" from Honolulu to Guam that made regular stops at Johnston, a service that helped make the atoll into the most visited of all the islands of the Pacific marine monument.

The smaller of the two "revised" islands was Sand Island, originally a roughly circular expanse of ten acres, about the size of a residential lot in a well-to-do American subdivision. It was refashioned mainly by acquiring a slightly smaller sister island, with a causeway constructed to connect them. In its final form, the island duplex was in the shape of a dumbbell, oriented in an east-west direction.

Still, these modifications were not quite sufficient to fulfill the total array of purposes that the military had in mind for the atoll. Thus, in 1964 a corps of engineers and manual laborers brought into existence two entirely new islands: North Island (otherwise known as Akau, the Hawaiian word for "north"), twenty-seven acres, and East Island (Hikina), seventeen acres, each a short distance from Sand Island.

All these transfigurations, amplifications, and island creations ex nihilo made the atoll into an all-purpose, multifunctional base of activities. In real estate terns, the atoll had been "rebuilt to suit." And over the years, the military put the atoll to an astonishing range of purposes. It functioned sequentially as a naval refueling depot, a submarine base, a seaplane base, an airfield, a missile base, a launch site and data collection post for atmospheric nuclear testing, a Coast Guard LORAN transmitting station, and the staging point for a series of biological weapons trials in the open ocean 150 miles to the west. Later, Johnston Island was made into a chemical weapons storage site. Later still, it became an incineration facility for chemical warfare agents leftover from World War II, the Vietnam War, and other conflicts, including lethal and incapacitating substances such as sarin, VX, mustard gas, vomiting agent, and Agent Orange, as well as other agents imported to the atoll from US chemical munitions stockpiles in West Germany.

It was an irony of the situation that during all of these different utilizations, the atoll was—of all things—a federal bird refuge. Its status as such went back to Alexander Wetmore's trip to Johnston in 1923 aboard the *Whippoorwill*. As a result of a memorandum later submitted by Wetmore on July 29, 1926, President Calvin Coolidge issued Executive Order No. 4467: "It is hereby ordered that two small islands known as Johnston Island and Sand Island, located in the Pacific Ocean . . . are hereby reserved and set apart for the use of the Department of Agriculture as a refuge and breeding ground for native birds."

The atoll thereby became known as the Johnston Island Reservation (later renamed the Johnston Atoll National Wildlife Refuge). The motivation for proclaiming the atoll a refuge was clearly to protect the birds, for the executive order made it "unlawful for any person to hunt, trap, capture, willfully disturb or kill any bird of any kind whatever, or take the eggs of such bird within the limits of this reserve, except under such rules and regulations as may be published by the Secretary of Agriculture."

Executive Order No. 4467 remained in full force and effect even though the atoll would come under the successive jurisdictions of the Department of the Navy, the Air Force, the Atomic Energy Commission, and other agencies, and despite the use of Johnston Island as a base for atmospheric nuclear weapons tests, which by any measure was an unusual activity to pursue on, above, or anywhere near a wildlife preserve.

The reason why the POBSP established a long-term banding operation there was to gather data that would allow the army to determine whether Johnston Atoll, or the open ocean grid nearby (the fixed Johnston Atoll Grid), would be a safe place to conduct biological weapons trials. The answer would be a function of whether this isolated area was heavily trafficked by birds flying to regions where they could bring the pathogens used in the trials and thereby infect people. If it turned out that Johnston Atoll *was* a major transit point for birds migrating to distant points elsewhere, then it would be an unsuitable location to perform open-air releases of disease-causing biological organisms. But if the atoll was not in fact a hub for long-distance avian flights to other places, then the chances of birds acquiring a pathogen during biological testing work and bringing it to distant human populations would be relatively low.

So, ostensibly, the army needed to know exactly what the situation was regarding bird flight to and from the Johnston area. However, there is persuasive evidence to show that the army had identified the atoll as the staging point for two of its most important biological weapons trials—involving the tularemia and Q fever microbes—well before the POBSP scientists finished their banding work and came to a conclusion about the degree to which Johnston Atoll was significantly transited by migrating birds.

In the end, Johnston Atoll became one of the most radically transformed, heavily utilized, and intensively studied group of islands in the entire Pacific. Quite possibly, more has happened per unit area on this tiny collection of land masses than has befallen any other place of comparable size—one square mile—on the face of the earth.

Indeed, perhaps even too much has happened here, at this artificial atoll of the Pacific.

* * *

Two names have been associated with the first sightings of Johnston Atoll by Westerners. One is Joseph Pierpont, captain of the Boston-based whaling ship *Sally*, which grounded on the outer reef on September 2, 1796. The place was named, however, for Charles James Johnston, who as captain of the British ship HMS *Cornwallis*, sighted the islands on December 14, 1807.

A later encounter by a member of the Wilkes Expedition (i.e., the "US Ex Ex"), recorded the reef's position and added, "On the northwest side of this reef there are two low islets: the one to the westward was covered with bushes, but no trees; the other was no more than a sand bank," which may explain the name of Sand Island. Neither island had a natural source of fresh water, and there was no evidence that either had ever been inhabited.

The issue of the atoll's ownership extended back to the guano period and was marked by an almost comic succession of claims, counterclaims, and the sale of land rights by purported owners whose legal possession of the islands or their guano reserves was highly dubious to begin with. Ultimately, the matter of rightful claim to the atoll was not settled until the United States annexed Hawaii in 1898 during the Spanish-American War, whereupon the atoll became a US possession.

It was during the years leading up to World War II that Johnston Atoll started undergoing its long series of physical changes and also began to be inhabited on a regular basis. Because of its location, Johnston Atoll was considered a natural stopover and refueling spot for planes flying from Honolulu to points south and west. And on account of its larger size, the US Navy contemplated making Johnston Island into an airport. By 1944 the navy had built a 6,100-foot-long runway on Johnston Island, which originally was only about one-third of that length. The addition of ancillary structures such as a control tower, taxiways, roads, military housing, recreation areas, a hospital, and the like completed the transformation of the atoll from a tiny, desolate hazard to shipping into a fully functional and well-outfitted military installation. At its peak, Johnston Island would have up to a thousand human inhabitants, most of them US Air Force personnel.

But the atoll was soon more than just an airport. In 1947, the US Atomic Energy Commission designated three areas of the Pacific Ocean as component sites of the Pacific Proving Grounds. One of them was the all-purpose, multifunctional, ever-useful Johnston Atoll.

On July 31, 1958, the United States launched the first of a series of high-altitude nuclear tests from Johnston Island, an event code-named Teak. A Los Alamos Laboratory scientific report on the environmental impact of the Teak blast stated, "After the event, we observed quite a few birds sitting or hopping on [Johnston Island] docks in a helpless manner. Either they had been blinded or they were unable to dive for fish, their major food supply, because the ethereal oils which protect their feathers from getting water-soaked had been boiled off by the thermal pulse."

There was to be another high-altitude nuclear weapons test, code-named Orange, with an even higher expected yield, twelve days later. On this occasion, the test planners took some protective measures for the birds. "An artificial smoke screen was generated to cover Sand Island at explosion time," the Los Alamos report said. "But Orange was fired above a rather dense cloud cover—perhaps fortunate for the birds escaping from the smoke."

The canonical POBSP report on Johnston Atoll provided further detail about precautions in place during subsequent testing: "An elaborate water sprinkler system was installed on the original portion of Sand Island to protect the birds living there. In addition, other protective devices were used, including smoke pots placed upwind as a shade screen and aerial flares to divert the birds' attention from the flash of the blast itself."

One of the most remarkable features of the artificial atoll was the dogged persistence of birdlife on all four of its islands despite the repeated waves of disruptions to their ecologies. At least half a million seabirds—most of them sooty terns, but including more than fifty other species as well—used the atoll for roosting and nesting. This is explained by the fact that the atoll is the only land available for breeding in the 450,000 square miles of open ocean surrounding it.

* * *

It was in 1963, barely six months into the original contract with the Smithsonian, when the army made an "urgent request" that the POBSP establish, among other things, a long-term and intensive bird migration surveillance program at Johnston Atoll. A proposal described the new work: "In July, 1963, a small field party will land briefly on Sand Island to undertake a preliminary survey. Up to now Sand Island has not been studied under the original contract. It is now proposed to undertake intensive ecological study of Sand Island more or less continuously using three to five men depending on personnel availability and logistics problems. The effort will include banding and migration and distribution analyses of birds as well as studies of other vertebrates. Because of the availability of regular air transport to Johnston Island it is expected that problems of logistic support for this program will be minimal."

It is a reasonable assumption that the army's "urgent" interest in the banding, migration, and distribution of birds to and from Johnston Atoll was a consequence of its realization that the fixed grid to the southwest was the ideal location for its contemplated large-scale biological weapons trials. After all, the military already controlled the area; Johnston Island had previously hosted other kinds of weapons tests, including nuclear detonations; and it was in the middle of nowhere. It was practically made to order for a biological weapons testing program.

Leonard Carmichael, secretary of the Smithsonian, signed the revised contract on June 24, 1963, and two weeks later, on July 7, 1963, the first two POBSP fieldworkers arrived. They were Binion Amerson and Kenneth J. Wilz, and they arrived aboard a Military Air Transport Service flight from Honolulu to Johnston Island. Ultimately, they moved to Sand Island, where they took up residence.

Their new living quarters were nothing like the beach party encampment lifestyle adopted on the islands that were the more usual haunts of the POBSP. There was at that time a US Coast Guard station on Sand Island, and Amerson and Wilz were lodged in the chief petty officer's quarters. This was a venue that offered all the comforts of home, with occasional steak dinners, cookouts, and free drinks. There was even a movie theater on the island, and one night they screened *Bird Man of Alcatraz*.

Sand Island was also home to a LORAN transmitting tower that was 625 feet high. This was a very tall structure, taller than the Washington Monument and fully half the height of the Empire State Building. Binion

Amerson was given the chance to climb the open-air tower but declined. Someone else accepted the same offer, climbed up, and took pictures from a dizzying height (fig. 5.2).

The tower was held securely in place by a system of twenty-five guy wires that stretched from the top of the structure to concrete anchors set in a circle in the lagoon surrounding the island. In addition, three other sets of cables ran from lower levels of the tower to points below on the island itself.

The guy wire complex constituted an obstacle course for the island's birds to navigate in flight to and from their nests, for the birds used the island for breeding. There were about two hundred thousand sooty terns nesting on the island, as well as birds of several other species. Many made their nests on the concrete blocks into which the wires were set.

Living there came at a price, however, for the birds in flight often collided with the cables, and this took a substantial death toll. Collision with the wires killed some two dozen sooty terns per day at the start of the breeding season when the birds swarmed above the island in a cacophonous, ear-splitting frenzy.

Fig. 5.2 View from LORAN tower on Sand–Johnston Island, including antenna supports, 1964
Credit: Smithsonian Institution Archives, RU 245 Box 225, Folder 11

It was on this first tour of duty on the atoll that Amerson and Wilz were joined by officials from Fort Detrick and the Deseret Test Center. The purpose of their visit was to assist with the collection and shipment of two hundred live wedge-tailed shearwaters, blood samples, and other specimens destined for delivery to Detrick and Dugway. Perhaps as a result of Fred Sibley's earlier, somewhat chaotic experience with the collection and shipment of 150 live Laysan albatrosses from Midway Island to Washington, this time the military would be on hand to ensure that the whole process would unfold more smoothly.

Not that much more smoothly, as it turned out.

"After dark we began processing Shearwaters," Binion Amerson recorded. "Ken caught the birds and held them down while I took the blood samples. I then killed the bird and gave it to Chief Gragosian. He obtained the spleen from each bird. We processed thirty before stopping at 12:30. We did all our work on a wooden crate."

The spleen is involved in the immune system's response to pathogens, and it appeared to be the army's hope that analysis of its contents would reveal what viruses, bacteria, or other agents the bird might have been exposed to.

The men processed thirty-two more shearwaters the next night. Meanwhile, the shipping boxes for the live birds arrived on a Military Air Transport Service (MATS) C-124 Globemaster cargo plane.

"Put twenty-five boxes together for shipping birds," Amerson wrote. "Discovered that the tape S. I. sent to be used is absolutely useless. It will not stick and is apparently old." They got some new tape the next day, put the boxes together, and labeled them.

"After dark Ken, John [Bushman], and I proceeded to the East end [of Sand Island] to capture Wedgetailed Shearwaters. Just as we finished it started to rain and we rushed them back to the trailer and headed back to the Pump House."

On August 22, 1963, finally, they got rid of both their living and frozen cargoes.

The final shearwater count was two hundred live specimens shipped, one hundred collected and killed for blood and spleen samples. There was an estimated total of approximately two thousand shearwaters on Sand Island, and so a loss of three hundred birds represented a decline of more than 10 percent of the population.

Hard times on the bird refuge.

* * *

The POBSP report *The Natural History of Johnston Atoll, Central Pacific Ocean* is one of the principal scientific documents to emerge from the Pacific Project. It was the joint production of four coauthors: A. Binion Amerson, Jr., Philip C. Shelton, Roger B. Clapp, and William O. Wirtz II. Two of them, Amerson and Shelton, had spent long stretches of time on the atoll, the former having stayed a cumulative total of 162 days there. But it was Philip Shelton who was the record holder for participation in bird studies at Johnston Atoll, having spent a cumulative total of 458 days—a year and three months—on Sand Island.

The report the four authors produced ran to 502 pages of text, charts, tables, diagrams, illustrations, and references, and it took up the entire December 1976 issue of the *Atoll Research Bulletin*. It lavishly illustrated the fact that one of the principal purposes of the POBSP's work at Johnston Atoll was to establish a record of seabird migration patterns to and from the area. And this record in turn was founded upon all the bird banding that the fieldworkers had done over the years.

Across their more than six years spent on the islands of Johnston Atoll, the scientists banded a total of 303,901 individual birds—an immense number. As for recoveries, there were two main ways of obtaining bird recovery data during the Pacific Project. The first was by members of the field teams recapturing banded birds themselves. By the end of the program, 60,932 birds had been retrieved by field team workers at Johnston Atoll. Of this total, 60,526 birds were originally banded on the atoll itself, while only 406 had been banded elsewhere—a very small number of birds given the total number of birds that had been recaptured.

The second source of recovery information was amateur bird watchers and professional ornithologists in other countries. In 1965, in an effort to obtain such data from these private sources, Roger Clapp started a popular newsletter called the *Pacific Bird Observer*. Subtitled, "Newsletter of the Pacific Ocean Biological Survey Program," its logo pictured a stylized bird that was probably meant to be a sooty tern, with a band on its right leg and a ribbon streamer on its left. The newsletter's editor was Tina Abbott, who would later become Clapp's wife (and later still, ex-wife).

The bimonthly newsletter was printed and published by the Smithsonian Institution's Division of Birds, which sent copies to various bird-watching societies and individuals to enlist their aid in recapturing banded birds and sending back location data. This was a modest success, and the newsletter printed the names and affiliation, if any, of some of the birdwatchers who recaptured banded birds at far-flung locations, mostly in the Northwestern Pacific.

From these and other sources, by the end of the program in 1969 the POBSP made an important discovery about bird migration patterns in the Central Pacific. A tabulation of banded seabird recoveries showed two main areas in which populations were concentrated in the Central Pacific Ocean. One was in the Northwestern Hawaiian Islands, including Johnston Atoll. The other consisted of the Line and Phoenix Islands, two separate island groupings near the equator. The recoveries showed that seabirds tended to congregate within each of the two areas, whereas there was not much migration between the two.

"The overall POBSP banding and recapture program in the central Pacific has shown that bird movement between the Line and Phoenix Islands area and the Hawaiian and Johnston area was virtually nil," the authors said following the end of the study. "Movements between islands within each of these areas was much larger, and the number of birds returning to their island of banding was still greater."

In other words, the birds tended to return to the same local area, rather than travel the long distance between the two island groups. But then came this surprise: "Banded birds from both areas, however, were recaptured in the far western Pacific."

This was a major problem insofar as it related to the army's biological testing program: the army had already conducted its two most important and potentially hazardous field trials in the open ocean near Johnston Atoll (the Shady Grove series of tests) in the months of February, March, and April 1965, long before any significant amount of bird recapture data had been generated either by the POBSP field teams or by the volunteers reached by the *Pacific Bird Observer*. Indeed, the Shady Grove tests were conducted while two POBSP field team members, David Bratley and Norman Heryford, were present on Sand Island and still assiduously banding birds.

But if birds from the Johnston area had flown through clouds of the active biological warfare agents used in Shady Grove, then conceivably they could have picked up and carried some of those agents to human populations in East Asia. And that possibility raised a crucial question: Had the army learned of this westward seabird migration pattern by early 1965, *before* it held the weapons trials in the area west of Johnston, or only afterward?

In answering this question it is important to note that the field teams on Sand Island did not make recapturing birds a primary focus of their activities until 1967, fully *two years after* the Shady Grove field trials were completed. In addition, the first issue of the *Pacific Bird Observer* newsletter was dated September 1965, *five months after* the field trials were over and done with. At the time of the Shady Grove tests, therefore, the army apparently had little or no access to the conclusions the scientists would later reach about bird migration patterns because the bulk of the data on which those conclusions were based had not yet been collected. The full dataset would not be available until long after the Shady Grove trials had been run.

It is true that the army knew of the *presence* of birds within or near the planned test area prior to the Shady Grove tests, for the POBSP report explained:

> In August 1963, a 50,000 square mile, rectangular, pelagic grid was established by the POBSP, centered approximately 175 miles southwest of Johnston Atoll [the fixed grid]. In all, 42 monthly survey cruises were conducted in this grid through February 1967. Birds, primarily seabirds, were observed along 22,898 miles of daytime travel and along 10,819 miles of night travel. Pelagic observations maintained for more than 2,500 hours in daytime and more than 1,150 hours at night recorded 33,261 birds of 41 species.[1]

This series of cruises went under the name of the STAR BRITE study, which was a specialized POBSP subprogram. But although the STAR BRITE observational study established the *presence* of birds within the grid area, it did not establish where those birds had come from or what their destinations were.

Still, it is possible that the army received at least a preliminary accounting of the POBSP's initial bird migration findings starting in 1963, when the first bandings and recaptures were made by field team members. The army could have received this information in the POBSP's semimonthly progress reports together with occasional notes that Binion Amerson and others sent to the army's Deseret Test Center starting in July 1963 and continuing for an indefinite period afterward. And the army might have concluded that this preliminary data was sufficient for its purposes.

But if on the basis of the preliminary data the army was aware that birds frequented the test area and traveled to the Northwest Pacific, why did it nonetheless conduct open ocean trials in the grid area using live hot agents when doing so might have negative health consequences among human populations there? Roger Clapp, for one, argued that the army should *not* have done so. In a 1998 interview he claimed that the Johnston Atoll area was the last place it should have used for biological weapons testing.

"They tested there because they ruled the roost at Johnston," Clapp said. "If there's anything we had shown, it's that you shouldn't mess around with biological agents near Johnston. You could use frigate birds to kill off everybody in the Western Pacific if you wanted to."

Why then, if Clapp is correct, did the army nevertheless disperse the Q fever and tularemia pathogens in the fixed grid area to the southwest of Johnston Atoll starting in February 1965? Indeed, if it was going to ignore its findings or conduct tests before full information about migration patterns was available, then why did the army fund the POBSP study to begin with?

Note

1. Night observations of birds in flight were made by radar. A November 1963 report by the Smithsonian's Division of Birds specified that one of the POBSP's objectives was to determine the "night-time status of birds by analysis of camera film recordings of continuous (24-hour) radar scannings of the sea and comparison with information obtained visually."

6

Project 112

The army brought the Pacific Ocean Biological Survey Program (POBSP) into existence and funded the project for its entire duration, for the simple reason that it was instructed to do so by officials higher up the chain of command within, and outside of, the US military. The military being what it was and is, when you receive an order, you follow it. In this case there was both a proximate mover and an ultimate mover of the relevant chain of events. The proximate cause was the secretary of defense, but the individual who was ultimately responsible for the Smithsonian's Pacific Program was the president of the United States, John F. Kennedy.

The story begins with Kennedy, who was inaugurated on January 20, 1961, the year East Germany put up the Berlin Wall. As often happens with an incoming administration, top officials wanted to take a bold new look at the status quo. That meant, among other things, launching a review of the nation's existing military readiness with the idea of making possible improvements if, when, and where necessary. At the height of the Cold War, Kennedy was intent on lessening the nation's reliance upon nuclear weaponry and upon the principle of mutual assured destruction, and he wanted to replace or supplement them with other, less risky, and more limited means of waging warfare. Chemical and biological weapons systems were two possible alternatives to nuclear bombs, and each option warranted a fresh study. Ironically, the fact that chemical and biological agents were potentially *less* destructive than nuclear weapons made them more suitable for waging war.

A few weeks after taking office, Kennedy appointed Robert S. McNamara, previously the head of the Ford Motor Company, as his secretary of defense. Early in his tenure, McNamara established a senior interagency task group to assess how the Pentagon was organized and how the branches of the armed forces were structured and equipped. The task group undertook a broad and sweeping policy review that resulted in approximately 150 sequentially numbered projects, each of which defined a study of one or more specific problem areas within the military.

Some of the projects were quite ambitious, others less so, and many were primarily organizational in nature and scope. Project 80, for example, was entitled Study of the Functions, Organization, and Procedures of the Department of the Army. Project 100, by contrast, would answer the narrow question of whether it would be more efficient to have a single manager of supplies for all three branches of service, or whether multiple distributed managers could accomplish the same functions more effectively.

Further down the list was Project 112. It was not organizational; it was substantive. In essence, its task was to evaluate the prospects of chemical and biological weapons for both strategic use as well as for employment in more limited, tactical warfare situations. And so in May 1961, McNamara asked the Joint Chiefs of Staff to undertake such a project, the fulfillment of which would turn out to be no trifling matter.

"Consider all possible applications, including use as an alternative to nuclear weapons," McNamara told the Joint Chiefs. "Prepare a plan for development of an adequate biological and chemical deterrent capability, to include cost estimates, and an appraisal of the domestic and international political consequences."

It would be an understatement to call this a tall order. Satisfying all the requests embodied in McNamara's capsule description of Project 112 would entail the creation of an entire new military command and a major study of the effects of a range of chemical and biological weapons. The Joint Chiefs thought that such weapons "had great potential" but that "they could be considered operational only in the most limited sense and that the task of measuring their impact accurately still had to be done."

Doing so would in the end require a substantial new weapons testing program. The necessary tests would have to be planned and coordinated, and their results evaluated and understood. The tests themselves would have to be conducted by an agency or group that was properly equipped for the task. And, of course, suitable test sites would have to be found. Given that these would be large-scale weapons trials, covering a wide geographical area, the sites in question would have to be located well outside of the continental United States.

Getting all of these functions operational involved the formation of three distinct entities. The first was a new US Army command post, the Deseret Test Center (DTC), in Utah. It would plan, schedule, and organize the tests, and it would evaluate their results. The second was Project SHAD (Shipboard Hazard and Defense). This was the fittingly opaque cover name for the mini–joint army, navy, marine corps, and air force strike team that would actually perform the tests, both chemical and biological. And the third entity was the POBSP, whose job it was to undertake a systematic search for one or more sites where the large-scale biological and chemical weapons trials could be conducted secretly and safely.

All of this was made necessary by Robert McNamara's otherwise innocuous-sounding Project 112.

* * *

Officials at the Deseret Test Center oversaw every aspect of both the SHAD test program and the Pacific Project. A heavily redacted, declassified version of an originally secret document describing the biological weapons trials that were conducted in the open ocean area southwest of Johnston Atoll furnishes a wealth of detail regarding the overriding requirement that the test be done "safely," that is, with no harm done to animals, plants, or people. The relevant tests were performed between February and April 1965, and the report documenting them was dated December 1966. It was called *Test 64-4 Shady Grove: Final Report* and was written by Ernest H. Buhlman of the Deseret Test Center. The document is unique in being a secret army report that contains an authoritative, internal, and proprietary account of the army's rationale and motivation behind the POBSP. It is further noteworthy in that it places the responsibility for the final selection of a test site not on the military itself but instead on a newly created group of civilian advisors, a body of medical and public health experts enlisted by the Deseret Test Center to give their professional advice as to the possible health effects of the contemplated biological weapon trials.

The preface to the document begins with a statement outlining the operational parameters governing the tests and the conditions under which they were to be performed:

The following guidance concerning pathogenic biological field testing was received from responsible authorities [not identified] about the time Deseret Test Center was organized in May 1962: (1) the tests were to be conducted with a minimum of equipment, support personnel, facilities, and elapse of time on site; (2) test sites were limited to United States possessions or remote open sea areas; (3) the dissemination of agent materials was to have no protracted or significant effect on the environment—this included people, domestic and wild animals, birds, or any biological life which might be permanently injured or could create a hazard to man; and (4), tests were to be conducted safely and in accordance with a security plan which would ensure a minimum of risk of detection.

The document then described the problem of site selection:

The first task was selection of a test site which would meet the criteria for causing no significant effect on the environment. The selection of the test site would define the extent of the problem which had to be faced in order to fulfill the other criteria.

To solve the problem, the Deseret Test Center assembled a medical advisory committee whose members would propose one or more test sites based on the information available to them at the time.

A Medical Advisory Committee was formed composed of eminent scientists from the field of ecology, epidemiology, and related sciences. The Chairman of the Committee was Dr. Dorland J. Davis, M.D., Director, National Institute of Allergy and Infectious Diseases, National Institutes of Health. This Committee first met in July 1962, to review information on a number of proposed test sites in the Pacific area with reference to the release of agents *Pasteurella tularensis* (LE), *Coxiella burnetti* (OU), and *Venezuelan equine encephalomyelitis* (NU).

The remainder of the committee was made up of experts in a variety of medical specialties, including an ecological officer from the US Public Health Service in Fort Collins, Colorado; a veterinarian from the Agricultural Research Service in Hyattsville, Maryland; the chief of the epidemiology branch of the Communicable Disease Center (CDC) in Atlanta, Georgia; and a medical entomologist from the Rocky Mountain Laboratory in Hamilton, Montana. Two other members were from Yale and the University of Wisconsin. Decidedly, this was an elite group.

The DTC Medical Advisory Committee was charged with estimating not only possible threats to human health arising from the release of pathogenic microorganisms into the open air but also their likely effects upon the health of the bird and mammal populations in the vicinity of the tests, as well as their consequences for the environment more generally. The committee members would recommend suitable locations for the biological weapons trials so that they would be as environmentally friendly, benign, and "green" as possible.

It is significant that the committee's first meeting was in July 1962, three months before the start of the Smithsonian's Pacific Ocean Biological Survey Program, and a year before the first POBSP field team, consisting of Binion Amerson and Kenneth J. Wilz, landed on Johnston Atoll. This means that the committee was being asked to make site selection recommendations well before its members had sufficient empirical field data on which to base reliable, evidence-based conclusions. The members proposed a site nevertheless and suggested a means by which they could make a more rationally informed choice.

> While there was insufficient information available at the time on which to base a final conclusion, an open-sea site was suggested as being the most acceptable. A program was outlined to develop the necessary ecological and epidemiological information on which to base a final decision.

The program in question was an Ecology and Epidemiology [E&E] Study, which army documents identify as the de facto genesis of both the POBSP itself and of a special-focus study that was a subsidiary of it. A separate, unclassified army report acknowledges unambiguously that the motivation for the POBSP came directly from the DTC Medical Advisory Committee:

> Acting on the recommendation of the Deseret Test Center Medical Advisory Committee, Deseret Test Center, from its establishment in 1962 to its merger with Dugway Proving Ground in 1968, sponsored a contractual E&E effort with the Smithsonian Institute [sic] and the University of Oklahoma. These programs provided required E&E surveys in those areas outside the continental United States which had been designated for possible open-air BW testing.

The Smithsonian's ecological work was embodied in two programs: the main program was the POBSP itself; the second was the POBSP subproject called STAR BRITE, mentioned previously. As described by the Shady Grove final report:

> A specialized ecology survey named STAR BRITE was started in July, 1963 as an intensive study of an open-ocean area (about $50,000_m{}^2$) southwest of Johnston Atoll, to evaluate and analyze the pelagic bird composition and

distribution and related factors that affect their ecological patterns over the proposed grid site for SHADY GROVE.

This means that in July 1963, before the DTC Medical Advisory Committee made its final recommendation on a proposed test site, the army had already defined the fixed grid area southwest of Johnston Atoll as a possible, desirable, and perhaps even preferred location for the Shady Grove tests. In March 1964, finally, the DTC advisory committee met again.

During the 21 March 1964 meeting of the DTC Medical Advisory Committee, all aspects of the Ecology and Epidemiology program (including a visit to Johnston Atoll) were presented by various support groups doing both field and laboratory work. The comprehensive reports and briefings were reviewed in detail by the committee.

On the basis of the committee's recommendation, the Deseret Test Center gave final approval to the Johnston Atoll open-ocean grid area as the test site for the release into the air of the *Pasteurella tularensis* and *Coxiella burnetti* pathogens in operation Shady Grove.

The selection of a site in the open sea southwest of Johnston Island seemed logical in view of the Medical Committee's suggestion; and since the site was located away from shipping lanes, it provided the greatest distance of downwind travel free of populated land areas anywhere in the Central Pacific and was known to have a low wildlife population.

And indeed, in February 1965, about a year after the pivotal March 1964 DTC advisory committee meeting, the army began the Shady Grove trials at sea.

This, then, was the basic progression of events:

(1) President John F. Kennedy appointed Robert McNamara secretary of defense.
(2) In May 1961, McNamara authorized Project 112.
(3) Implementing that project led to the creation of the Deseret Test Center, which brought together a civilian medical advisory committee to recommend one or more possible locations for biological weapons tests.

(4) The committee met in July 1962 and suggested that an ecological and epidemiological study be performed to gather the data to make an informed decision about test sites.

(5) In the late summer or fall of 1962, in an effort to get the required ecological study under way, three Deseret Test Center officials came to Washington and met with Remington Kellogg at the Smithsonian.

(6) This led to the contract signed by Leonard Carmichael in October 1962, making the Smithsonian a subcontractor to the US Army Biological Laboratories at Fort Detrick, and formally establishing the POBSP.

(7) Eight months later, in May 1963, the army made an "urgent request" that the Smithsonian widen the scope of the Pacific Project to include an intensive and continuous ecological study of Sand Island, Johnston Atoll.

(8) Shortly thereafter, in July 1963, the army authorized the STAR BRITE study, a specialized twenty-four-hour biological assessment of the pelagic bird composition and distribution across the open-ocean grid area southwest of Johnston Atoll.

(9) In March 1964, in light of the initial data provided by STAR BRITE study together with the equally preliminary data from the Sand Island ecological survey, the DTC committee suggested that the army's designated grid area would in fact be suitable for the Shady Grove tests of the tularemia and Q fever pathogenic agents.

(10) And in February 1965, Project SHAD's joint army, navy, Marine Corps, and Air Force strike team embarked on operation Shady Grove in the fixed grid area southwest of Johnston Atoll.

A convoluted tale, but one based on the army's own documentation of the procedures it followed to get from initial directive (Project 112) to final fulfillment (Shady Grove).

* * *

A question left unanswered by this sequence of events is that of the origin of the safety requirement to begin with. It goes without saying that a biological weapons trial should do no harm to human health, but why should that same requirement be extended to bird, mammal, and even plant life?

There are at least two answers to this. The obvious answer is that the safety requirement was extended to birds because they could pose indirect risks to human health by picking up a pathogen, carrying it elsewhere, and precipitating disease outbreaks in areas beyond the test site. This was what Roger Clapp meant when he said, "You could use frigate birds to kill off everybody in the Western Pacific if you wanted to." To prevent that from happening, whether by frigate birds or any other species, the army imposed the doctrine by which a proposed site was deemed "safe" if the acquisition and foreign transmission by birds of the biological agent in use during the trials was unlikely. A limitation of this answer is that it does not account for the inclusion of plants, which do not normally acquire pathogenic agents and migrate.

But there is a second answer, which is complementary and supplementary to the first: the army was also motivated by a general sense of ecological awareness, by a desire to protect the environment and living things in general, including plant life. Now, it might be assumed that the US Army was more or less immune to such essentially aesthetic or moralistic impulses. But that assumption would be wrong.

Despite the fact that the military is often perceived as insensitive to the damage caused by weapons research and development—the case of atmospheric nuclear testing being perhaps the best illustration—there is a way in which the army was nevertheless very much at the forefront of environmental awareness and particularly of the ecological effects of its activities. The modern environmental movement came into national prominence during the 1960s, in part as a result of the impact of Rachel Carson's *Silent Spring*, which was published in 1962. (Coincidentally, this was the same year the DTC Medical Advisory Committee was formed and first met.) Carson's book described the effects of certain pesticides, mainly DDT, on birds and is widely regarded as having given birth to the environmental movement in the United States. But the surprising fact is that regarding environmental issues, the army was well ahead of Rachel Carson. The army had become environmentally conscious, ecologically forward looking, and "green" more than a decade before the rise of environmentalism in the general culture.

In 1951, at the order of the chief chemical officer of the US Army Chemical Corps, Dugway Proving Ground instituted a program to inventory and study the plant and animal life within its borders as well as in the areas immediately adjacent. The goal was to amass baseline data on the plant and animal distributions, population dynamics, and the general ecology of the region. To assist in this effort, in 1952 Dugway entered into a contract with the University of Utah to undertake an E&E survey. Under the terms of the contract, specialists from the university were charged with a number of tasks, the first of which was to identify the plant and animal species living in the designated area and to develop a reference collection for the positive identification of a species of interest. A second task was to establish the potential for transmission of the candidate biological warfare agents being contemplated for testing by vectors (such as ticks or other biting insects) that were naturally resident on the wildlife in and around the proving ground. The University of Utah's E&E program was in effect a land-based template for what the Pacific Project would become some ten years later.

It is worth noting that "potential for transmission" is a loaded term, because knowledge of an agent's transmissibility could be of interest to the military for two entirely different reasons. One is *protective* in nature. If a given biological agent is known or thought to be easily transmissible to other locations and of causing disease among humans, then sufficient reason exists to avoid disseminating that agent in open-air tests where such transmission is likely. Here the knowledge of transmissibility is used in a proactive and precautionary manner, to prevent and protect against the introduction and spread of disease.

But this same knowledge is equally applicable to *offensive* use, as part of a deliberate attempt to *cause* disease in an enemy population. Indeed, that is the whole point and purpose of a biological warfare program to begin with. These two motivations are not mutually exclusive, and in the end, the army used its knowledge of transmissibility for both purposes. While it generally attempted to avoid inadvertent disease transmission, there was yet an entire division of the US biological warfare project that studied the use of *insects* as disease vectors. The army had long ago developed an *entomological* warfare program, first at Fort Detrick and then at Pine Bluff Arsenal, in Arkansas (see appendix II). And some critics have charged that while the ostensible motive behind the POBSP's study of bird migration patterns was to avoid inadvertent transmission, the army's real motivation was to surreptitiously investigate the prospects of intentionally using birds as vectors of disease, just as it had done with its entomological warfare program years earlier.

Whatever the plausibility of that contention, the environmental and ecology studies done by the University of Utah in 1952 were in the first case primarily descriptive: the investigators simply documented what was out there. But a second component of the program was to establish a rationale for minimizing the risk presented by the testing of human, animal, and plant pathogens (including *Bacillus anthracis*, the anthrax microbe) at Dugway.

In an attempt to provide the basis for an informed recommendation about the safety of testing such agents, in 1953 the Department of Defense convened the Ad Hoc Committee for Dugway Proving Ground. The committee, with both civilian and military members, was chaired by the US surgeon general at the time, Leonard A. Scheele, MD. In meetings at both Fort Detrick and Dugway, the committee formulated what has since come to be a generally accepted precept of US biological agent testing. Basically, the doctrine stated that those agents already present in the United States in animal reservoirs, and were relatively widespread, were deemed safe for testing at the proving grounds. The underlying premise seems to have been that if a given agent is already endemic in substantial quantities to begin with, then no additional harm would be done by introducing yet more of it. As a corollary to this, it followed that if an area could be found where bird and animal life was nonexistent or relatively scarce, then testing of an infectious agent in that area was correspondingly safe.

However, where lack of sufficient evidence of endemicity of a given pathogen existed, then simulants (such as *Bacillus globigii*, the anthrax stand-in), should be used in place of live, "hot," agents. The Scheele committee further recommended that "to reduce even such a small hazard as might develop, continuous surveillance of the rodent and ectoparasite populations should be continued."

The same operational policies that pertained to testing in areas within the United States also pertained to tests outside it, so when the Deseret Test Center was created in 1962, the army imposed upon it and its activities the same general E&E safety requirements formulated by the Scheele committee in 1953 and in effect at Dugway thereafter. And in turn, when the DTC gave rise to and sponsored the Smithsonian's Pacific Program, its field teams operated under the same set of safety requirements and surveillance procedures. In effect, therefore, the POBSP that came into existence in 1962 had its origin, roots, and rationale in the army's own environmental and ecology program that Dugway's Ad Hoc Committee had originated in the early 1950s.

* * *

It is ironic that for all the emphasis the army placed on safety, and on protecting human and animal life and the environment, there yet existed an escape clause, a sort of get-out-of-jail free card, to bypass all of its own safety requirements and to allow potentially hazardous large-scale biological tests to proceed anyway, irrespective of their possible adverse consequences. This escape clause came in the form of National Security Action Memorandum No. 235, dated April 17, 1963, issued by President Kennedy.

The short, two-page memorandum was entitled, "Large-Scale Scientific or Technological Experiments with Possible Adverse Environmental Effects." These effects included, according to the text, "significant or protracted effects on the physical and biological environment." The document then proceeded to outline a simple series of steps to circumvent the general prohibition against such tests.

First, any agency proposing to conduct tests that might have significant or protracted effects on the environment had to bring its proposals to the attention of the special assistant to the president for science and technology.

Second, "in support of proposals for such experiments, the sponsoring agency will prepare for the Special Assistant for Science and Technology a detailed evaluation of the importance of the particular experiment and the possible direct or indirect effects that might be associated with it."

Third, the special assistant will then review the supporting materials and make a recommendation about whether to proceed with the experiment.

After that, it was up to the president. "Any experiment that may involve significant or adverse effects will not be conducted without my prior approval," Kennedy stated. That is, the president could at his discretion authorize a test or experiment no matter what its possible adverse effects might be.

As it happened, the Shady Grove tests took place only after the then-sitting president, Lyndon Baines Johnson, signed off on the test series. This suggests (but does not by itself prove) that the army thought that these tests were in fact potentially unsafe but considered them to be important enough to national security to warrant their being conducted regardless. If this was in fact what happened, then it would have rendered all the work done by the STAR BRITE study, and by the ecological survey of Sand Island, useless, unnecessary, and beside the point, at least concerning its military value. Its scientific value, on the other hand, was quite another matter.

* * *

An important assumption underlying Roger Clapp's claim about killing large populations by birds infected with lethal pathogens, and also underlying the army's support of the Pacific Program, was that birds could transmit acquired diseases to humans relatively easily, reliably, and efficiently. And that assumption is not without foundation: it is widely known, for example, that wild aquatic birds constitute the main reservoir of influenza A viruses, that migrating ducks disseminate influenza viruses worldwide, and that those viruses can be transmitted to domestic poultry and mammals, including humans. And the birds in question have been quite successful at this.

Further, pathogens can be transmitted from birds to humans indirectly, from contact with water contaminated by waterfowl feces, and through vectors that are carried by wild birds such as mosquitoes and ticks. For example, the microbial cause of Lyme disease, *Borrelia burgdorferi*, can be transmitted by ticks that are carried by several common species of birds, including robins, cardinals, and sparrows.

It is a separate question, however, whether wild and migratory birds play a *significant* role in the direct or indirect transmission of pathogens, especially when carried by birds across long distances. If migratory birds do *not* in fact play a significant role in the transmission of a pathogen, then the scenario Clapp envisioned, and indeed the prospect of infecting large populations by birds that have picked up pathogens from the testing of biological warfare agents, becomes far less probable and worrisome.

In 2008, an international group of biomedical scientists published a major review study, "Human Infections Associated with Wild Birds," in the *Journal of Infection*. The impetus for the study came from the then-recent spread of diseases such as the avian influenza A and West Nile viruses. The fact that diseases caused by those viruses were observed in areas far from the locations where they were first identified generated "the hypothesis that migratory bird transported these pathogens to new geographical locations," the authors said. "However," they added, "as is the case with the highly pathogenic avian influenza, scientific data do not always support such hypotheses." Avian influenza was not in fact transmitted by migratory birds but rather by the movement of infected poultry through the normal routes of human commerce and trade.

To assess the actual impact of infections caused by migratory wild birds, the authors performed a literature search that ranged over 168 published articles about the transmissibility of disease by birds. They divided transmissibility into three categories: *direct* transmissibility if there was evidence of transmission through direct physical contact between birds and humans; *indirect* if the transmission was through insect vectors such as mosquitoes or ticks, or from water contaminated by the infected bird species' droppings. Finally, there was a *theoretical risk of transmission* when a pathogen was isolated from both humans and wild birds but no reports of either direct or indirect transmission. In other words, the link in such cases was merely cor-relative rather than causative.

The group's first finding was that direct transmission was very rare. "The literature review identified no real evidence for direct wild bird to human transmission with the only exception being the cluster of H5N1 human cases in Azerbaijan where the affected patients were plucking feathers from mute swans that had succumbed to H5N1 infection."

There was better, although still very limited, evidence for indirect trans-mission. The authors found a total of fifty-eight bacterial, fungal, or viral pathogens for which wild birds could serve as reservoirs, mechanical vectors, or both. However, "scarce microbiological, serological and epide-miological data supported indirect transmission from wild birds to humans for 10 of these pathogens." Examples of indirect transmission included *E. coli* bacteria, salmonella, and the spirochete *Borrelia burgdorferi*. There was only a theoretical risk of transmission for the forty-eight pathogens that remained.

The researchers found that the transmissibility of pathogens by wild birds over long-distance migrations was indeed possible. Once infected, the birds could carry the disease to humans by several routes. "Generation of con-taminated aerosols by waterfowl flocks may result in respiratory infections through the inhalation of dust or fine water droplets generated from infected bird feces or respiratory secretions in the environment." Food-borne infections may result from consumption of undercooked meat or organs of infected wild birds. "Infections may lastly result after direct contact with the skin, feathers, external lesions or droppings of infected wild birds."

An additional finding was that the pathogens causing disease may be transported over long distances by birds carrying infected ticks. "During mi-gration, there is sufficient time for some birds to travel hundreds or even a few thousand miles before ticks complete feeding and drop off," the authors wrote. "Even if these birds have small tick burdens, their large numbers could result in substantial contributions to local tick populations in coastal areas. There is even evidence of transhemispheric exchange of spirochete-infested ticks by seabirds indicating the capacity for wild birds to carry infected ticks for long distances."

In sum, the evidence reviewed by the authors "suggests that many pathogens can infect multiple host bird species and that these pathogens could in theory could be responsible for emerging infectious disease outbreaks in humans and wildlife." But there was this caveat: "The available evidence suggests wild birds play a limited role in human infectious diseases." There are simply far more common ways people acquire infections than by association with birds: namely, by person-to-person transmission, by eating infected meat sold commercially, and by other routes.

The takeaway from all this is somewhat equivocal. Although birds can and do carry pathogens over long distances, and can cause disease outbreaks in human populations, they do so on a limited basis. Roger Clapp's sweeping assertion must therefore be viewed as hyperbolic. While it would be theoretically and empirically possible for frigate birds to carry disease-causing pathogens to foreign populations, the prospect of entirely wiping out those populations by such means is highly improbable.

* * *

The scientists working on the Pacific Project remained blissfully unaware of the complex matters of military policymaking taking place over their heads, beyond their knowledge, and behind the scenes. Back on the islands themselves, field team members continued to do their bandings, take blood samples, perform bird counts, and produce bird skins. While at sea between islands, some of them devoted part of their leisure time to reflecting on what it was they were doing and how well they were doing it. Despite his youth and his reputation of being something of a class clown, one of the most thoughtful and ruminative of the scientists was Larry Huber, who recorded his impressions in a handwritten journal. On January26, 1965, shortly after leaving Honolulu on Southern Island Cruise No. 7, he wrote, as if confessing to his diary:

> I am getting screwed already this trip—No excelsior. Nobody back at the SI (except people who don't know anything about stuffing big birds) likes cotton bodies. I can't put up a good skin with a cotton body so this next batch of birds shouldn't help my standing at all. If Carl in taxidermy ever finds me doing sea birds with cotton bodies he'll never help me with birds again. Bober will never talk to me again. Arrgh. I hope we don't collect many large birds this trip. Since I've got to skin a lot of the birds I wish someone would let me have a say as to what I skin with. With all the tons of sawdust available to the project I find 100 lb of cornmeal aboard for skinning. No wonder the project is always complaining about its birds being no good. Maybe some day it will dawn that top quality birds can't be put up with poor quality materials.

Later on in the same cruise, while on Jarvis Island, Huber wrote about the cat infestation there and about the difficulty of eradicating them. Cats killed birds, and for that reason Larry Huber hated them.

The cats have adapted to our method of hunting them. They are no longer afraid of us as they used to be. They are now afraid of lights, motion and sound. This means the days of the shotgun on Jarvis are over. We have to counteradapt. This is war. If we don't eliminate the cats this next trip all the shotgun shells, time and effort will have been for nothing. The first thing that ATF [Auxiliary Tugboat Fleet] has to do is steal the 22 rifle from At Sea. Then have a 4X scope installed on it. Then when we get to the island two people go out and hunt for cats together. One with a 12 volt hand light and the other with the 22. Aim between the two glowing eyes and chalk one cat off the list. This is the only way to get rid of the cats on Jarvis. ATF has to have an accurate long distance gun.

None of the field teams that landed on Jarvis Island managed to eradicate the cats. That feat, which was harder to accomplish than it appeared, would not be done until much later, in 1990, by people who had nothing to do with the POBSP.

At sea again on the same trip, which also took him to Howland, Baker, McKean, Hull, Birnie, Enderbury, Canton, Samoa, and Christmas Islands, as well as to Palmyra Atoll, Larry Huber wrote:

There is something all wrong here. The general trend on this project is for quantity instead of quality. This in itself isn't so strange since it is a temporary project. What bothers me is that after a very elaborate explanation that because this is a temporary project as much information as possible has to be gotten out of the specimens, therefore the specimens do not have to be good but useable so that they can be studied now instead of in the future. So I start putting up useable skins so that they can be studied in the present. But then I hear nothing but complaints about how bad the skins look. . . . So I start putting up good skins which of course slows me down. [On an earlier trip, Huber recorded that "I find Frigatebirds very easy to skin. When I get going I do one every twenty minutes."] This doesn't satisfy anybody either because not enough skins are put up.

Indeed, precisely because it was a temporary project, funded by the army for a given period, and then later renewed for another year, successively, field team members commonly assumed that as the end of a given funding period approached, they would soon be out of a job. As Southern Island Cruise No. 7 neared its end, therefore, Huber thought that his participation in the project itself would be over as well, and he would soon return to the University of Arizona to get his master's degree. In the end, however, in 1965 the army renewed the contract for another year, and Huber came back and sailed with another Southern Island Cruise, in August and September 1966. In fact, he stayed with the program through 1969. But he didn't know that at the time, and as trip number 7 was steaming back to Honolulu in March 1965, Huber wrote:

> Somehow I am going to miss all this. Next September I'll be in college. No more thousands of birds, no more Hawaii and Samoa, no more petrels, no nothing. After this project I'll never be satisfied with those drab, nondescript land birds. . . . I'll never get to see the Laysan or the Bonin or the Sooty Storm petrel. So goes the days of wild surf and sharks while skindiving. . . . I may reach 80,000 [banded birds] but never 100,000. [On cruise No. 7 alone Huber had banded a grand total of 13,432 birds.] Time and rules will regain their position in my mind. How am I going to make the transition back to that dull life back in those dull states. . . .
>
> Here I am, two months on an ATF trip, 16 hours from port and I am not happy to be getting back.

7

"Bird Bombs"

The army's original contract with the Smithsonian Institution expired on October 14, 1964. At that time Leonard Carmichael, who had put his signature on the document, was no longer secretary of the Smithsonian. He retired on January 31, 1964, and was replaced by S. Dillon Ripley, who took office the following day. As head of the institution, Ripley now bore ultimate responsibility for the Smithsonian's continued acceptance of US Army funds for the Pacific Ocean Biological Survey Program (POBSP) and for the four separate contract renewals that would extend the program's fieldwork until June 30, 1969.

As Carmichael had prior connections to the CIA, so Ripley had prior connections to its predecessor agency, the OSS, or Office of Strategic Services. In view of their shared backgrounds in the spy business, it would be hard to find two other men more positively disposed to the idea of linking the Smithsonian to a secret military program.

Ripley was a man of great knowledge and experience in ornithology and was very much a hands-on administrator. Although he was no stranger to hardship, the sufferings he had endured were of his own making, largely the result of his having contracted malaria and other diseases while trekking through remote tropical jungles or mountain highlands in search of exotic birds. He was born into Eastern style, wealth, comfort, and privilege, and spent parts of his youth and later life on the Ripley estate in Litchfield, Connecticut, site of the family's sixty-one-room mansion. The building was named Kilvarock, after a castle in the Scottish Highlands.

Dillon Ripley was a man to whom successive waves of good fortune routinely arrived unbidden. He would receive unsought-for job offers and unexpected opportunities for free luxury travel to far-flung destinations. He had never been forced to compete in a Darwinian "struggle for existence." Far from it.

His father, Louis Arthur Ripley, was a stockbroker and real estate entrepreneur. His mother, Constance, showed him the world. At age ten, she took him to Paris, where he sailed model boats on the pond in the Tuileries. On their further tour of Europe, they stopped at London and Vienna, as well as at selected spots in Scotland and Italy.

The young Dillon developed an interest in birds at the Fay School, a private boarding school in Southborough, Massachusetts, where his teachers led him on bird walks through the sixty-six-acre campus. He became a bird lover in consequence, and as a senior at Yale he decided to become an ornithologist. But before he could pursue the appropriate graduate work at Harvard, his life was interrupted by a surprise proposal that arrived in the mail. A well-to-do couple, the Denison Crocketts, who were friends of the Ripleys, were planning to sail their fifty-nine-foot schooner, the *Chiva*, to Dutch New Guinea. "We need a zoologist and wonder if you would care to go," they asked Dillon.

This was like Charles Darwin being asked to sail for Galapagos aboard the *Beagle*. Irresistible, but it would mean an indefinite postponement of his graduate studies. But Dillon knew Ernst Mayr, the curator of birds at the American Museum of Natural History, in New York, and asked his advice. "You cannot get to New Guinea every day," Mayr told Ripley. "Graduate work will still be waiting when you come back and then, too, you will be better prepared for it."

What came to be known as the Denison-Crockett Expedition sailed from Philadelphia on December 1, 1936, and it took them ten months just to get to New Guinea, a sea voyage of 12,000 miles. When Dillon arrived back in Boston a year and a half later, on July 14, 1938, he brought with him thousands of bird skins plus a collection of live birds in cages.

Ripley entered the netherworld of espionage during World War II, by means of his network of connections at Yale, whose faculty and graduates were being recruited by the office of Wild Bill Donovan, soon to be head of the OSS. Ripley was invited to join Donovan's staff on the very day the Office of Strategic Services was founded, June 13, 1942.

After the war, in 1946, Ripley was fairly inundated with job offers: a faculty position at Harvard, an associate curatorship at the Smithsonian, and an appointment at Yale, which was just an hour from the family estate. He chose Yale, which offered him a dual position as assistant professor of zoology and associate curator at the Peabody Museum, which was in effect a dinosaur preserve.

After eighteen years at Peabody, where he rose to director, he had acquired so great of a reputation as a curator, leader, and promoter of museums that when Leonard Carmichael retired as secretary, the search committee formed to find a replacement put Ripley's name into immediate contention. As it was, he did not even have to come to Washington for an interview; instead, a member of the search committee came to him in New Haven. And of course offered him the job, which he accepted.

Things fell into Dillon Ripley's lap so easily and so regularly, as a matter of course, that when, during his early years as secretary, the army offered to renew the POBSP contract for yet more folding cash money, he was not one to say no. "To me, as a bird man, this was a wonderful breakthrough," he said later. "It was a source of funds. That's all I know about it."

* * *

The contract extension that Philip Humphrey negotiated with the army in the summer of 1963 required not only that the Smithsonian fieldworkers establish an intensive and continuous bird surveillance program at Sand Island in Johnston Atoll but also that they would visit, among other places, the Leeward Islands northwest of Hawaii, making "brief biological surveys of as many islands as possible several times a year." Some of the "islands" in the Northwest Hawaiian chain were the smallest and most insignificant specks of land imaginable. Disappearing Island, for example, at six acres in total size and all of eight feet in height was periodically awash during rough seas, and it was invisible in heavy rains. And there were other land masses that, although they harbored bird populations, weren't even "islands" in the true and proper sense of the word.

La Pérouse Pinnacle, for instance, is less an island than a mountaintop or stone promontory that just happened to have risen from the sea, seemingly out of nowhere and for no apparent reason. It was part of the French Frigate Shoals, a crescent-shaped group of small land masses approximately 550 miles northwest of Honolulu. The pinnacle was 122 feet tall at its highest point, with a secondary peak a short distance away, and a saddle-like depression between them. From a distance, this configuration gave the prominence the somewhat startling appearance of a full-rigged sailing vessel plowing through the seas. An even smaller outcropping, known as Little La Pérouse, all of nine feet high, lay four hundred feet to the west.

The French explorer Jean-François de Galaup, Comte de la Pérouse, in the early morning hours of November 6, 1786, almost wrecked his ship, the *Broussole*, on the rocks, but lookouts sighted the whitecaps of the surrounding breakers just in time for the craft to be brought about. In 1788, during a later stage of his round-the-word expedition, La Pérouse vanished completely into the Pacific near the island of Vanikoro in the Solomons.

Despite its being difficult to land on, and hazardous to climb because of its crags, steep faces, loose rocks, and guano deposits all over the place, the POBSP scientists made several landings on La Pérouse Pinnacle over the years—and even more landing *attempts*. A common theme in the field notes pertaining to La Pérouse was a statement to this effect: "Due to the large swells breaking over the rocks and the sheer cliffs we were unable to land. We headed back for the ship and got thoroughly soaking wet since we were headed directly into the wind and waves."

But starting in September 1964, Pacific Project fieldworkers managed to land there successfully and then to climb up stepwise from ledge to ledge to points from which they could make inventories of several bird species, including ruddy turnstones, wandering tattlers, and masked boobies. The men returned to La Pérouse on several occasions between 1965 and 1969 and updated their observations. No birds are known to have been banded, nor specimens taken, alive or dead, from La Pérouse Pinnacle (fig. 7.1).

But as small as it was, and as difficult of access, La Pérouse Pinnacle was not the most challenging of the islands facing the Smithsonian field teams. That

Fig. 7.1 La Pérouse Pinnacle, French Frigate Shoals, Northwestern Hawaiian Islands

Credit: Dr. James P. McVey, NOAA Sea Grant Program

distinction belonged to Gardner Pinnacles, about 140 miles to the northwest of La Pérouse.

Gardner Pinnacles consisted of two volcanic rock outcroppings: a larger one rising some 150 feet above sea level and a smaller peak about thirty yards away that rose to a height of about 100 feet, together comprising a total land area of approximately five acres. As Roger Clapp described them in his report: "Barren, of no commercial value, and difficult to land on, these pinnacles are the least visited islands of the chain."

Nevertheless, visits were made, including one by Alexander Wetmore in 1923 and two by members of the POBSP in 1963 and 1967. Despite their commercial uselessness, barrenness, and cosmic insignificance, the US military found them logistically important, as their location made them valuable as navigational aids and mapping outposts. To create a flat spot for a landing area for helicopters, a military explosives team simply blew the summit off the pinnacle. As Clapp described it, "The larger of these peaks was formerly about 170 feet high but blasting away of the top reduced its height somewhat. This was done in March 1961."

Because of its near-vertical rock faces on all sides, only two landings were made on the island over the course of the almost one hundred years between 1828 and 1923. The latter was by Alexander Wetmore, who stopped there as part of the Tanager Expedition on the morning of May 22, 1923. The landing itself required some time and talent, but once ashore climbing the rock face was not impossible. "Though the sides of the rock were steep they were eroded out so that progress was only a matter of climbing up over series of ledges," Wetmore wrote in his journal. "The summit was gained without particular difficulty."

The first POBSP field team, consisting of Fred Sibley and Binion Amerson, arrived at Gardner Pinnacles on June 16, 1963. They landed by whale boat at the southeastern tip of the larger island at 8:00 in the morning and spent the next seven hours observing, cataloging, and collecting specimens of some the island's life forms, everything down to and including beetles, flies, and ticks. The island supported only scant vegetation, just some low-growing, mossy ground cover known as *Portulaca*, which was found in isolated spots on the slopes.

Of the larger animals, the scientists noted a single reptile, a green turtle, swimming offshore, and two monk seals sunning themselves on the smaller of the rocks. The seals were on a ledge about two feet above the water line, and they had gotten there by riding the sea swells until they were level with the surface.

The rocks were tailor-made for birds, however, and indeed there existed a great profusion of them, of fifteen different species, on the two pinnacles. There were as many as a thousand sooty terns alone, plus an estimated five thousand brown noddies, together with lesser numbers of blue-faced boobies, and others. In a mere seven hours spent on the island, Sibley and Amerson banded 416 birds of all fifteen species. The pinnacles were also home to many kinds of insects, including spiders, earwigs, and ticks.

What the army made of Gardner Pinnacles is unknown.

* * *

We do know, however, what the army was making of the live specimens the POBSP fieldworkers captured and sent back to Detrick and Dugway. Specifically, they turned the birds into test subjects and performed experiments on them. However, according to some later critics of the Pacific Project, the army's true purpose in these experiments was to establish whether the birds could be made into carriers of infectious diseases, in other words, into pathogenic agent delivery systems. The army's biological warriors, these critics said, intended to utilize gulls "for a doomsday assignment." They were trying to create "bird bombs."

On the face of it, this interpretation of what the army was up to in its support of the POBSP was not wholly implausible. It has been known for centuries that animals were carriers or vectors of diseases that are transmissible to human beings. The Great Plague of the Middle Ages, after all, was caused by a bacterium harbored by rat fleas. At Fort Detrick, the army had already researched the use of mosquitoes as biological agent delivery systems and established a mosquito production facility at Pine Bluff Arsenal for the purpose of raising millions of the bugs. And as we have seen, it was well known that birds, especially wild ducks, were carriers of the influenza virus, which causes a human disease that is sometimes fatal.

So what was so outlandish about the concept of making birds into weapons? Allegedly, there was even a "smoking gun" in the form of a Fort Detrick technical report that went back to October 1963, during the very time that the POBSP scientists were routinely making their Southern Island Cruises and shipping planeloads of caged birds from Johnston Atoll to Honolulu, and then onward to the mainland, to Detrick and Dugway.

And at first glance, the document in question indeed appears incriminating enough. For this is Fort Detrick Technical Manuscript 99, "Susceptibility of Sooty Terns to Venezuelan Equine Encephalitis (VEE) Virus." Its first paragraph baldly states, "A portion of the work reported here was performed under Project 4B92-02-034, 'BW Agent Process Research.'"

BW Agent Process Research? What could be more damning than that?

In nature, the Venezuelan equine encephalitis virus is transmitted to humans by mosquitoes that feed on horses or other equines suffering from the disease. Human patients typically develop a prostrating syndrome of chills, high fever, headache, and body aches. The disease is rarely fatal in adults but has a significant fatality rate among school-age children. At Detrick, scientists had long investigated VEE as a possible biological warfare agent. If someone wanted to use sooty terns to carry it, the first order of business would be to establish that birds of that species were susceptible to VEE infection to begin with. And that was precisely the object of the experiment described by Detrick's Technical Manuscript 99.

Sooty terns are one of nature's stranger anomalies. They are medium-size birds, about sixteen inches long and with wingspans almost a yard wide. They are the most abundant seabird breeding in the Central Pacific and the most common bird species in the Northwestern Hawaiian Islands. Sooty terns spend most of their time in the air, sometimes several years without alighting on land, and even sleeping while in flight, using the biological equivalent of an aircraft autopilot system. They come ashore only to nest and breed. But, despite being a truly oceanic species, sooty terns do not spend any time on the water (alighting only rarely on floating objects), because their feathers lack the oils to make them waterproof. Thus, as if in an evolutionary lapse, accident, or cosmic joke, sooty terns do not float.

But it was their very profusion across the Pacific that made them of interest to a cohort of Fort Detrick's biological warfare scientists. Technical Manuscript 99, issued in October 1963, was written by four members of Detrick's Technical Evaluation Division, and said in part: "The birds employed in this study were sooty terns (*Sterna fuscata oahuensis*). Two hundred and fifty-seven of the species were captured in the wild state and provided to these laboratories for testing." Although the report does not say where they came from, it is likely that the birds had been taken and shipped from Johnston Atoll by members of the POBSP.

Many of the birds were in poor shape upon arrival, and the report noted that "because of large numbers of nonspecific deaths during the testing period, only 135 were maintained for a sufficient period to provide test data." This means that of the original 257 shipped, some 122 of the birds were dead on arrival. The survivors were kept in wooden, wire-front cages in a large room and were fed a diet of chopped squid twice a day.

The VEE virus used in the experiment was grown in eggs at Detrick, and to establish that it was infectious, the scientists first injected it into twenty-four "Swiss-Webster mice," eight of them at each of three different levels of viral concentration. All the animals died within ten days of the inoculation, showing that the virus was indeed a lethal strain, at least to the mice.

For sooty terns, the Detrick experimenters varied the pathogen dose levels and generated an abundance of data about the responses to the virus at each successive level. In the primary experiment, the terns were subjected to an aerosol cloud of VEE virus particles suspended in a cloud chamber. This was a somewhat grisly process.

"Birds were exposed to aerosols by inserting their heads into the cloud chamber through slits in rubber diaphragms mounted in the chamber wall," the report said. "During the exposure the birds' bodies were confined in small boxes." Such boxed animals—including rhesus monkeys, guinea pigs, rats, and mice—were a common fixture of Fort Detrick aerosol testing work.

The results showed that although the birds developed the viral infection, they did not die from it or even manifest "obvious signs of illness," except at extremely high and unrealistic dose levels. In general, therefore, the report concluded that "VEE virus infections are not fatal for sooty terns. . . . It seems likely, then, that normal living habits would not be seriously disrupted during an infection."

It is noteworthy that there is nothing in their report to suggest that Fort Detrick scientists were trying to create a "bird bomb" or an avian doomsday device with these experiments. The most natural and reasonable interpretation of these tests is that they were merely the army's attempt to satisfy the overriding ecological mandate of Project 112 that large-scale scientific biological warfare trials should do no damage to the environment or to the life forms within it. What these results showed is that the use of VEE was consistent with that standard since the birds were not adversely affected by the virus. It would be safe, therefore, for the army to perform large-scale biological agent trials using VEE in an area of the Pacific where sooty terns were common. The birds would be infected by the virus, but nevertheless would not be harmed by it.

But if the experiment showed that sooty terns were not adversely affected by VEE virus aerosols, then why did the army never perform any such large-scale VEE trials in the Pacific, as it initially intended to do? The answer is that a *second* part of the experiment showed that the dissemination of such an agent posed a possible risk to *human* health. In the second phase of the experiment, mosquitoes were allowed to feed on the infected terns while their blood serum was at high VEE viremia levels. After their blood meal, the mosquitoes, now carrying the virus themselves, were allowed to feed on healthy, uninfected sooty terns. And when they did so, the previously healthy birds developed VEE infection as well.

That in turn meant that mosquitoes carrying the virus from infected birds to healthy birds could equally well transmit the VEE virus to human beings and cause infections among them as a secondary and unintended consequence. This provided sufficient reason not to disseminate VEE virus aerosols in the Pacific. And indeed, in the large-scale biological agent trials that the army later conducted there, the VEE virus was never used.

Two further pieces of evidence suggest that these experiments were not designed to create a sooty tern doomsday device. The first is that sooty terns were above all *pelagic* birds, living primarily in oceanic environments. Thus, they have no natural propensity for flying to cities that would be likely targets of a biological warfare attack. They are not, like pigeons, rife in public parks, playgrounds, and on the sidewalks of major metropolitan areas. Being an oceanic species, they are radically unsuited for the role of bringing infectious diseases to human populations.

But whereas sooty terns have no propensity for flying into human population centers, mosquitoes certainly do. For a mosquito, a human being is a blood meal. Given this, and the further fact that the army already had an operational mosquito production facility at Pine Bluff Arsenal, it would make more sense for the scientists to use mosquitoes themselves as vectors of disease than it would sooty terns. Smithsonian bird banding returns showed that sooty terns flew where they pleased. Mosquitoes flew toward human beings.

A second bit of circumstantial evidence also points to the conclusion that Fort Detrick's Technical Manuscript 99 does not represent a scientific master plan for building a bird bomb. And this is that the report had been unclassified from the very start, whereas army documents describing biological warfare research, devices, or weapons systems were routinely classified as secret or confidential. Technical Manuscript 99, by contrast, was and is freely and easily available even today through a simple web search.

The experiments the report describes, therefore, were not efforts to turn birds into bioweapons. They were, rather, for the army, an exercise in due diligence.

* * *

A different scientific report, in this case originating at Dugway Proving Ground, provides further evidence as to what the Army actually did with the live bird and blood samples taken by the POBSP field teams. This is *Summary Report on the Susceptibility of Birds to Tularemia: The Wedge-tailed Shearwater and the Black-footed Albatross*. The document, written by Dugway researchers C. A. Brown and V. J. Cabelli, was dated October 1964. It was initially classified as Secret but was downgraded at twelve-year intervals until it automatically became unclassified.

The report describes the outcomes of tests performed at Dugway's Baker Laboratory. The tests had multiple objectives, among which were (1) to establish whether the species of birds in question harbored antibodies to the tularemia pathogen, (2) whether they could be experimentally infected by the pathogen, and (3) "to determine the extent to which *Pasteurella tularensis* can be transmitted from animal to animal by direct or indirect contact."

Pasteurella tularensis (a bacterium later renamed *Francisella tularensis*), was the causative agent of tularemia, also known as rabbit fever. In humans, this is an incapacitating disease characterized by localized skin ulceration, fever, and, occasionally, pneumonia. To the army, the microbe was attractive as a biological warfare agent because it was infective at extremely low doses, on the order of one to ten bacterial organisms per person. The experimenters knew from previous tests on two species of noddies and two species of terns that "all four species were highly susceptible to infection and the lethal consequences thereof." The new tests, on approximately a hundred birds brought back from the Pacific, found that "the Shearwater is fairly resistant to the lethal effects of the disease but is very susceptible to the infection." (The shearwaters in question were in all probability those captured on Sand Island by Binion Amerson and Ken Wilz and shipped from Johnston Island to the States in August 1963.)

As for the black-footed albatross, the Dugway scientists had a quantity of serum samples "taken from the birds 'on site' as evidence of past and present infection. No such evidence was obtained with some 250 samples examined

to date." This result suggested that the albatrosses had not contracted tularemia in the wild. Further, attempts to induce the disease among live albatrosses in the lab showed that "the Black-footed Albatross is relatively
insusceptible to the lethal effects of experimentally induced tularemia."

Finally, the experimenters found a low likelihood of transfer of the disease
from animal to animal in birds of either species. The final conclusion was
that "on the assumption of low transmission efficiency by the oral contact
route the potential transfer of tularemia through fecal discharge of these bird
species may be so low as to be insignificant."

It follows from this that both the black-footed albatross and wedge-tailed
shearwater were unsuitable candidates for use as "bird bombs." Indeed,
critics who allege that this was part of the army's purpose in its bird studies
have produced no actual evidence to the contrary.

All the same, it is difficult to reconcile everything that is reported in this
document, which was issued in October 1964, with the fact that in February
through April of the very next year, 1965, in operation Shady Grove, the army
did in fact disseminate the tularemia pathogen over a large area of the Pacific
near Johnston Atoll. For although the Dugway report established that both
the wedge-tailed shearwater and the black-footed albatross were not susceptible to the lethal effects of tularemia, it also stated that previous tests showed
that birds of four *other* species "were highly susceptible to infection and the
lethal consequences thereof."

This means that the army knew in advance that introduction of the tularemia pathogen *could* in fact have adverse consequences on the bird life in
the area of the planned large-scale sea trials but went ahead and conducted
the tests anyway. Why?

A possible answer may be derived from a further series of facts. First, recall
that the DTC's Medical Advisory Committee met on March 21, 1964, and
that during the meeting "all aspects of the Ecology and Epidemiology program (including a visit to Johnston Atoll) were presented by various support
groups doing both field and experimental laboratory work. The comprehensive reports and briefings were reviewed in detail by the committee." And, on
the basis of the committee's recommendation, the Deseret Test Center gave
final approval to the Johnston Atoll open-ocean grid area as the test site for
the release into the air of the *Pasteurella tularensis* (tularemia) and *Coxiella
burnetti* (Q fever) pathogens in the Shady Grove trials.

What is unclear from these accounts is whether at the time the Medical
Advisory Committee was aware of the Dugway studies of the susceptibility

to tularemia of the wedge-tailed shearwater and the black-footed albatross. Although the report was dated October 1964, the results were known to the Dugway experimenters well before publication (the tests were performed between January and June 1964). It is therefore possible that those results were communicated privately to the advisory committee at their March 1964 meeting.

It is also unclear when the susceptibility of the four other species (two species each of noddies and terns) was established, although the relevant Dugway report, on the "tularemia potential for transmission by birds," is dated August 1964. It is therefore likewise possible that the medical advisory committee was privately informed of these results as well during the March meeting.

Further information might have come to the advisory committee from the STAR BRITE program. That program, which started in July 1963, would eventually show that the sooty tern was the most common species in the grid area, with the wedge-tail shearwater and sooty shearwater less common. It is important to note that the STAR BRITE observations showed a clearly-defined seasonal pattern to these sightings: there was a low population density in the winter months, from December through March, but a high population density in the summer months. Further, there was a spring migration peak in April and May, and a fall migration peak from September through November.

These facts argued that if the Shady Grove tests were to be conducted at all, then they should be performed during the winter, the period of low population density. And if the committee members were so informed, then they might well have made a recommendation to that effect.

In addition to the STAR BRITE study, the army also carried out a separate epidemiology program during which researchers obtained serum samples from the grid area's wildlife and human sera "from five islands in French Polynesia which represent both rural and urban indigenous adult populations." Analysis of the serum samples revealed the presence of antibodies to both the tularemia and Q fever pathogens in some species of wildlife, as well as tularemia antibodies (but not Q fever antibodies) in the human samples. This meant that at least some wildlife had acquired a protective immunological defense against both pathogens, whereas humans had acquired such a defense to only one.

That lack of immunity to Q fever constituted a good and sufficient reason *against* using the Q fever microbe in the Shady Grove tests. But mysteriously enough, the army used it nevertheless. Indeed, given that four species of birds were known to be susceptible to the lethal effects of the tularemia microbe, and that human serum samples showed the absence of Q fever antibodies, the fact that the Shady Grove test disseminated both pathogens is the most difficult single feature of the case to understand or explain.

Any account of what the Medical Advisory Committee's recommendation might have been regarding the proposed Shady Grove tests is necessarily speculative, but some indicators point to what the committee was likely to have decided.

There were two possible reasons why the members of the advisory committee might have recommended that the tests proceed despite the possible threat of disease transmission to certain bird species and to human beings. One is that the tularemia pathogen was thought to be a nonpersistent agent that rapidly degraded in sunlight. Thus, it is likely that the committee would have recommended that if the agent was going to be released at all, it should be disseminated during the early morning hours, at a point when the pathogen would have sufficient time to infect a host, but then shortly afterward lose its virulence to ultraviolet irradiation. In this scenario, the pathogen would have only a limited time in which it was infectious, thereby minimizing the risk of its causing disease among birds and, later, people.

The Q fever microbe, however, had a spore form that was resistant to heat, ultraviolet irradiation, and sunlight. The committee would therefore most probably have recommended the release only of the non-sporulated form. (It is not in the public record which form of the microbe was actually disseminated in Shady Grove.)

In addition, the military routinely evaluated the meteorological conditions prevailing at the time of the tests, so that that any agent released would be blown out toward empty sea and away from land areas. The advisory committee may have recommended such a release as another means of reducing the potential risk of disease transmission to foreign lands.

What is known for sure is that when the Shady Grove biological warfare trials were conducted, they took place in the previously identified open-sea grid area southwest of Johnston Atoll. The tularemia agent was released between February 12 and March 15, 1965, during the winter months, when the POBSP scientists found there was a low bird population density in the fixed grid area. Later, between March 22 and April 3, 1965, the Q fever agent was also released in the same stretch of open ocean.

Finally, the army took meteorological conditions into account, and the Shady Grove tests did in fact take place in the early morning hours, just before and just after midnight of each day. The most logical inference then, is that the medical advisory committee recommended that the tests take place near midnight and during the winter months; that is, more or less as and when they in fact took place. Confining the tests within such time frames would allow the sea trials to proceed, while at the same time minimizing the risk of the Q fever microbe being transmitted elsewhere.

But all of this is retroactive speculation and is hypothetical at best. What the advisory committee actually said, and why, will probably never be known for sure. Until and unless new documents or other pertinent information comes to light, these unknowns will continue to be a part of the dark or invisible history of the American biological warfare program and of the POBSP's role in it.

8

The Military Payoff

Given how much it invested in the Pacific Ocean Biological Survey Program (POBSP) across seven years (an estimated $3 million), it is remarkable how little the army in fact derived from the program that it itself inaugurated and sponsored. In fulfilling the army's commission, Pacific Project field teams traveled all over the North and South Pacific: to the Northwest Hawaiian Islands, to the Line Islands, to the Phoenix Islands, to the Marshalls, to American Samoa, and even to the Aleutians. They banded 1.8 million birds and captured and shipped to the mainland literally hundreds of live and skinned bird specimens, plus countless blood, spleen, stomach contents, and other biological samples. The field teams generated massive, almost overwhelming amounts of data, on the basis of which its members issued well over a hundred preliminary reports, interim reports, final reports, and publications. But what the army is known to have done with this gigantic supply of information amounts to very little in the end.

As we have seen, the purpose of the POBSP was to find extra-continental locations suitable for large-scale chemical and biological weapons tests to be carried out under the auspices of Project SHAD (Shipboard Hazard and Defense). And indeed, the army did in fact conduct three separate tests of biological weapons systems in the years the POBSP was active: the Shady Grove and Magic Sword trials, both in 1965; and Speckled Start, in 1968. But of the three, only Magic Sword seems to have been a genuine product of Pacific Project field research. For that test the army needed an island free of mosquitoes, and members of the Pacific Project had discovered that Baker Island, where the Magic Sword trials were held, was indeed mosquito-free.

As for Shady Grove, it is very likely that the army had decided on the open-ocean fixed grid area near Johnston Atoll long before the DTC Medical Advisory Committee recommended it. To all appearances, therefore, the Deseret Test Center (DTC) committee seems merely to have rubber-stamped a location that the army decided upon well in advance. (The committee may also have suggested that the tests be performed at specific times of day, and in certain seasons.)

The third and final test, Speckled Start (DTC 68-50), was conducted at Eniwetok Atoll in the Marshall Islands. But Eniwetok had been visited by the POBSP on only one occasion, in June 1966, and then not again until August 1968, the month before the Speckled Start test was to take place. In 1966, furthermore, the sole field team member there (Dale Husted) did not even set foot on the island; inexplicably, he spent the whole of his twenty-four hours there aboard ship. It is doubtful, therefore, that the POBSP could have contributed any significant amount of useful information to the army on the basis of the single, one-day, offshore visit or the later visit just prior to the test.

Still, the POBSP and Project SHAD were very much intertwined. They were both created by Project 112; they ran concurrently, essentially from 1962 to 1969; and some of the vessels that transported the Smithsonian field team members from island to island were also used in the three tests. A full understanding of the POBSP and its effects therefore requires knowing what the objectives of the tests were, as well as how they were conducted.

The field trials themselves were elaborately planned, consequential military exercises from which the army drew important conclusions about the general utility, practicality, and effectiveness of biological weapons. These tests, and their results, constituted the military payoff of the Pacific Ocean Biological Survey Program.

Chronologically, the first of the three biological weapons trials was Shady Grove. Initially, the test had been called Red Beva, short for "research and development, biological evaluation." Ultimately, the Deseret Test Center's planners decided that the name gave away what the test was about (whereas its meaning was in fact quite opaque). The trials were therefore renamed Shady Grove, which had no hidden meaning attached to it, although it was the name of a residential neighborhood in Montgomery County, Maryland, just north of Washington, DC.

The Shady Grove trials involved the dissemination of two pathogenic biological agents: the tularemia microbe (military code UL), and *Coxiella burnetii*, the Q fever organism (OU). According to the declassified final report on the test, the objectives of the tularemia component were fourfold:

1) To evaluate infectivity of agent UL aerosols over effective downwind distances, utilizing an elevated line source from an operational weapon in a marine environment.
2) To determine the viability decay of UL over effective downwind distances.

3) To characterize atmospheric diffusion in a marine environment.
4) To assess the operational capability of the weapon system.

Operational headquarters of the trials was the USS *Granville S. Hall* (YAG-40), a converted Liberty ship assigned to the Deseret Test Center as part of its naval fleet. (YAG was the navy's designation for "miscellaneous auxiliary service craft.") The *Granville S. Hall* had seen previous duty in the Operation Castle series of nuclear tests at Bikini and Eniwetok Atolls in 1954. At that time, the ship was outfitted with instrumentation that allowed it to be operated by remote control by a small crew in a sealed hold, which meant that it could pass through radioactive fallout and measure its intensity without presenting any hazard to the crew. After completion of the fallout sampling, personnel clad in protective gear and clothing would decontaminate the ship.

Later, during the Pacific Project, the *Granville S. Hall* transported Smithsonian field teams on several Southern Island, Northern Grid, and Eastern Area Cruises. By this time, however, the *Granville S. Hall* was quite a different craft. It was now a floating biomedical laboratory. The ship's medical complex, a collection of small rooms forward of the superstructure, was equipped with gimbaled tables for keeping autopsy surfaces level even in heavy seas, plus analytical instrumentation, autoclaves, and other standard-issue medical lab equipment. The test subjects, rhesus monkeys, were housed in either of two separate holding areas: an open space for those that had not yet been exposed to the test agent, and a second, sealed and filtered hut called the "doghouse," for those that had. In effect, the *Granville S. Hall* was the seaborne equivalent of a Fort Detrick biological agent aerosol exposure facility.

The *Granville S. Hall*, its sister ship, the USS *George Eastman* (YAG-39), and the five Army light tugs used in the trials (LT-2080, 2081, 2085, 2086, and 2087) were "citadel" ships. This meant that there were special, enclosed areas of the interior that were ventilated through HEPA—high-efficiency particle arrestor—filters and were otherwise hermetically sealed off from the external environment. The crews on the ships were confined to the citadels for the duration of the aerial release of the hot agent and the ship's subsequent passage through the pathogen cloud.

The Shady Grove tularemia trials took place between February and March 1965. In charge of the operation was J. Clifton Spendlove, a civilian scientist working for the army at the Deseret Test Center as the technical director of the Plans and Evaluation Directorate. The light tugs, stationed in Honolulu, were under the command of Lt. Commander Jack Alderson of the US Navy and were ready for operations as of January 2. They had to wait, however, until President Lyndon Johnson signed the necessary authorization. "He signed it on 21 January 1965," Jack Alderson recalled later. "We were underway on 22 January, proceeding to Johnston Island."

The first aerial release of the tularemia pathogen occurred on February 12, 1965, in a portion of the remote fixed grid area whose center was about 175 miles southwest of Johnston Atoll. Prior to and during the conduct of each trial, three P2V Neptune aircraft overflew the area to look for accidental entries of unauthorized ships. From overhead, the operation was directed by a US Navy radar-equipped Lockheed Constellation aircraft, which acted as an airborne command post, provided air traffic control, and helped in positioning the light tug fleet.

Starting at twilight, navy crewmembers on the tugs set out the rhesus monkeys in three to six cages placed at various points on the deck. "A seasick monkey is not a pleasant thing to deal with," Jack Alderson said, "but they would be hung around the ship." After positioning the cages, the crew entered the citadel area until the test was over. During the exercise, the tugs maintained radio silence.

Four hours after midnight on February 12, 1965, a single Marine A-4C Skyhawk approached the grid area from the east. The Skyhawk was a single-seat, subsonic jet fighter, and it carried a set of Aero 14B spray tanks, one beneath each wing. The tanks were long and thin and had four fins attached to the tail section. At the "function time," which in this case was at 0402 hours, the jet began dispersing two agents simultaneously, the tularemia microbe from one tank, and a biological tracer microbe from the other, through a discharge nozzle at the aft end of each.

Biological tracer microbes were bacteria that when dispersed into the air left physical vestiges of their presence, concentration, and distribution. In the tularemia trial, the tracer was *Bacillus globigii* (BG), a bacterium commonly found in the environment. When the BG bacteria were deposited on Petri dishes containing a growth medium, they produced an identifiable, telltale, greenish pigment.

The jet released the two agents along a straight line that was anywhere from twenty to thirty-six miles in length, a few miles upwind of the light tugs. The tugs then steamed through the resulting aerosol cloud, sometimes more than once. Random birds might have done so as well, inasmuch as POBSP observers in the STAR BRITE program recorded radar sightings of birds in the grid area at night.

At 0433 on the same night, the jet made a second pass and laid down another line of the tularemia agent and the BG tracer. It then returned to the air base on Johnston Island.

In a 2002 congressional hearing on Project 112 and the SHAD tests, Jack Alderson, the tug fleet commander, described what happened next.

> At the morning twilight, and after the cloud had passed the tugs, then three sailors would go topside, pick up the Rhesus monkeys, take them down, and place them in the doghouse. They would pick up the samplers, the petri dishes, and so forth, and take those down to the doghouse, seal that up. They would then decon the exterior of the tug, and the tug would get underway, go alongside the Granville S. Hall. We would take our samples and our test subjects, pass them to the Granville S. Hall, and get new subjects and equipment for the next day, proceed back to the grid to the next assigned position.

The Shady Grove tularemia trials would continue until March 15, 1965, and consisted of twenty-three separate disseminations of the tularemia microbe. A week later, starting on March 22, essentially the same set of actors, ships, and aircraft performed releases of the second hot agent, *Coxiella burnetii* using the same procedures. Across the span of two weeks the Skyhawk jet made nine additional passes, each along a line about forty miles in length.

After the dispersals, Jack Alderson explained, the *Granville S. Hall* scientists "watched the reactions of the monkeys for a period of time, then they were euthanized, given autopsies and put in an autoclave, ground up, and put in the ocean."

But there is a codicil to the story. As recorded in the Deseret Test Center's final report on Shady Grove, "Ecological and epidemiological activities were increased during the conduct of the test and for two months following in an effort to detect if any change was made in the biological environment. The DTC Medical Advisory Committee met again 3 June 1965 to evaluate these data. There were no indications that any change had taken place."

Insofar as the US Army was concerned, therefore, and insofar as anyone else has been able to show, the Shady Grove tests were, in the end, "safe."

* * *

The second set of bioweapons trials, Magic Sword, was the single test series that was actually based on data supplied by the POBSP. But the data in question was only the simple fact that the island used in the trials was free of mosquitoes. This meant that any mosquitoes that wound up in the traps set by the experimenters were guaranteed to have been raised in the lab and were not endemic to the area.

Magic Sword began just two months after the conclusion of Shady Grove, and the two sets of tests could not have been more different. Everything about this new round of trials was radically unlike its predecessor. For one thing, the new trials involved no infectious agent at all. Instead of pathogens, what the military experimented with in this case was a *delivery system*, specifically the *Aedes aegypti* mosquito. In nature, the *Aedes aegypti* carries viruses that caused serious diseases, such as yellow and dengue fevers. But the mosquitoes used in Magic Sword were uninfected, disease-free, and wholly innocent. Indeed, these tests were so low-key, lacking in drama, and unthreatening to human life and health as to be almost boring. Nevertheless, the army conducted them with the utmost seriousness of purpose and practice.

Aside from the absence of pathogens, the location was also different, as the trials were held on Baker Island, near the equator.

A third difference was that no tugboat fleet was used in the trials, or any aircraft—only the laboratory ship *George Eastman*. A fourth was that the experimental subjects in Magic Sword were human beings, consisting of volunteers from the ship's crew. A fifth difference was that in addition to the military personnel that ordinarily conducted the tests, the Magic Sword trials included a Fort Detrick scientist who played an active role in the proceedings. This was Roger A. Scherff, who worked in Detrick's C (Crop) Division as a plant physiologist. More surprising yet was that a member of the POBSP itself was also present: Robert S. Standen was there to monitor the test's possible effects, if any, on the island's flora and fauna.

Magic Sword (which may have been named after the mosquito's biting apparatus) was part of the army's ongoing entomological warfare program. In the early 1950s, a subset of the army's biological warriors established an entomology division at Fort Detrick. While the entomological research proceeded there, the army also built a separate insect mass-production facility at Pine Bluff Arsenal.

In his book *Six-Legged Soldiers: Using Insects as Weapons of War*, Jeffrey A. Lockwood claimed that, "by 1960, the entomologists at Fort Detrick could produce half a million infected mosquitoes per month. As impressive as this was, it fell far short of the number of insects needed to attack a major city.

So the military drafted plans to increase production by a thousandfold." The plans were realized at Pine Bluff Arsenal, which "would house the largest insect-rearing rearing facility in the world, a mosquito mill with the capacity to produce 100 million infected vectors per week."

The Magic Sword trials evaluated the *Aedes* mosquito's biting habits and tested its ability to cross a stretch of seawater and remain viable enough to infect people. But as simple as it was in concept, in actual execution, a considerable bit of logistics was required.

The challenge was to transport the needed mosquitoes while keeping them alive and healthy over a distance that was literally thousands of miles away, to a hot, remote speck of land on the equator. Entomologists knew that mosquitoes could be kept viable for sustained periods of time if their environment was cooled slightly to reduce their metabolic rates. And so in May 1965, scientists at Pine Bluff placed a massive quantity of mosquitoes (precisely 1,644,500, according to one source) in an air-conditioned moving van, where they were held at a temperature of 64° Fahrenheit and in an atmosphere of 80 percent humidity to prevent dehydration. After thirty-six hours total elapsed time aboard flights from Arkansas to Honolulu, the mosquitoes needed to be taken still farther, to Baker Island, some 1,200 miles to the south. They made this stretch of the voyage aboard the *George Eastman*.

According to the Deseret Test Center's declassified report, *Test 65-4—MAGIC SWORD*, the experiment "was designed to study the feasibility of an offshore release of *Aedes aegypti* mosquitoes and to obtain additional information on (1) mosquito-biting habits, (2) mosquito trap technology, and (3) operational and logistical problems associated with the delivery of mosquitoes to remote sites." One such remote site was the coast of Vietnam, where the US Army might use mosquitoes as vectors to bring to that country diseases such as VEE, Venezuelan equine encephalitis.

And so in May 1965, the *George Eastman* arrived at Baker, "a low profile island of approximately 380 acres," and positioned itself three miles offshore. Volunteers from the crew, along with Detrick's Roger Scherff and POBSP's Robert Standen, tried to cross the distance on a landing craft, but the boat was swamped in the surf and sank. They tried again in rubber rafts, which worked. Once on the island, the experimenters set up an array of baited mosquito traps.

There were two types of mosquito releases during Magic Sword. The first was a release from a central point on the island, with human test subjects stationed in concentric circular arrays around the release point. The first such test was performed at a barren spot of land; all the rest took place within a vegetated area that more closely simulated the tropical conditions of coastal Vietnam.

It was Roger Scherff's job to propel the mosquitoes from a mechanical "mosquito release fixture," a canister-shaped device that forced a specific number of insects out the front end by a metered puff of gas. The tests started shortly after dawn and continued for six 10-minute intervals. The results of the trials—including both the number of mosquitoes found in traps and the number of bites received by each volunteer—were meticulously taken down and tabulated by the experimenters, whom one envisions as uniformed men roaming around the island with clipboards.

In the evenings, after the close of each day's set of releases, the test personnel eradicated any stray mosquitoes by spraying the island with an insecticide from a fog apparatus mounted on an "army mule," a four-wheel-drive, all-terrain vehicle.

The other kind of trial was an offshore release from the *George Eastman* itself, which at the "function time" was stationed three miles from the island. Oddly, there was only a single trial of this type.

At the end of it all, the men performed a final, island-wide mass extermination of mosquitoes with the fog generator. Still, the residual possibility remained that a few leftover, straggler mosquitoes might have secreted themselves somewhere aboard the vessel that brought them to Baker Island. To guard against that contingency the ship itself was cleared of errant bugs by a combination of insecticide and high heat.

That done, the USS *George Eastman* collected its navy crew members, plus the two scientists, Scherff and Standen, and steamed away.

* * *

There was only one additional US Army set of biological weapons trials at sea associated with the Pacific Project. It took place three years after the end of Magic Sword and was conducted on one of the least frequently visited and most hastily surveyed locations of the entire POBSP. But there was perhaps a good reason for this cursory treatment: the location in question was Eniwetok Atoll, which after a series of nuclear explosions there in the 1950s was already sufficiently well known (and, according to Roger Clapp, "was already royally screwed-up" by atom bombs) so as to require little additional exploration by the Smithsonian scientists.

Even so, the paucity of data collected by the POBSP about Eniwetok is remarkable given the otherwise exhaustive and repeated observations normally made by its personnel everywhere else. Moreover, when it was visited in 1966, the observations were made by a single field team member, Dayle N. Husted, who was on what must have been the loneliest voyage of the Pacific Project. For he was the sole POBSP scientist on Miscellaneous Pelagic Cruise No. 5, from Honolulu to Guam, aboard the US Coast Guard cutter *Basswood*, which was not a small vessel—180 feet in length.

The trip took almost a month, from June 2 to 27, 1966, sailing from and returning to Honolulu. En route, the ship made stops at Majuro, Kwajalein, and Eniwetok Atolls in the Marshall Islands, and at the US Naval Base Guam.

Husted's typewritten trip report on the cruise includes a single paragraph describing his visit to Eniwetok, an atoll on which he never set foot:

> We arrived at Eniwetok on the 21st of June. All observations here were in the lagoon because the observer [Husted] could not leave the ship. Very little bird activity was observed in the lagoon. Common Noddy Terns, Hawaiian Noddy Terns, Fairy Terns and a few light-phase Wedgetail Shearwaters were seen. Four Crested Terns were observed during the day. Just before sunset, 2 Golden Plovers were seen heading across the lagoon. At night, when we were anchored in the lagoon, several Black-naped Terns were heard and 2 were seen. These birds were not seen in the daytime lagoon observations.

And that was that. At least he knew his birds.

The entries pertaining to Husted's time on Guam were also somewhat peculiar. "Due to an oversight, the Project Observer had to receive shots and vaccination before he could leave the island. . . . Time on Guam was spent almost entirely on the Navy Base."

Still, it was easy to understand why the army would choose Eniwetok for a series of open-air trials of an infectious agent. First, the location was extremely far from the continental United States, meaning that whatever pathogen was dispersed there could present no possible hazard to Americans. The Native Marshallese Islanders were another matter, but even then, Eniwetok was so isolated and far to the west of the remainder of the Marshall Island group that the dissemination of a pathogen would pose little risk to them. Eniwetok is the most remote atoll within the Marshalls, so distant from the capital, Majuro, that it took three days by boat to get there.

Second, the nuclear explosions and the radioactive fallout to which Bikini and Eniwetok atolls had already been subjected meant that there was little additional damage that a dispersal of biological agents could do to the ecologies of their respective islands. The most extreme example of the environmental havoc wrought by the nuclear tests was what had happened to one of the islands that makes up Eniwetok Atoll, Elugelab.

Elugelab was a small diamond-shaped islet of forty acres on the northern perimeter of Eniwetok Atoll. The roster of species there consisted of the usual list of birds found on the islands of the southern Pacific: sooty terns, white terns, brown noddies, wedge-tailed shearwaters, and the like.

Unfortunately for the birds, Elugelab had been chosen by the US Atomic Energy Commission as the site of the world's first hydrogen bomb test. Around the time of the test, no native islanders lived on the atoll, only military and civilian personnel. And well before the event itself, all of them were evacuated.

The H-bomb detonation occurred at 7:15 on the morning of November 1, 1952, and within a matter of seconds, what had once been a peaceful tropical island vanished into thin air. In its place was now a mile-wide crater filled with water. Compared to that cataclysm, the dispersal of a few insubstantial lines of a pathogenic biological agent amounted to little more than a minor annoyance.

The new and final series of biological weapons tests, Speckled Start, DTC 68-50, took place during September and October 1968, near the end of both the POBSP and SHAD programs. These trials were in effect a replay of the Shady Grove tests near Johnston Atoll. They involved the same mother ship, the *Granville S. Hall*, and the same fleet of army light tugs as before. However, there were 100 fewer personnel (127) involved in this test series than in Shady Grove (223). The project manager of the new tests was Edgar "Bud" Larson, a Fort Detrick bacteriologist who had done early trials of the tularemia agent in Fort Detrick's enormous, million-liter aerosol chamber known as the "8-Ball."

The biological agent used at Eniwetok was SEB, or Staphylococcal enterotoxin, Type B, which was a toxin produced by the bacterium *Staphylococcus aureus*. According to the *Textbook of Military Medicine: Medical Aspects of Chemical and Biological Warfare*, "This toxin was especially attractive as a biological agent because much lower quantities were needed to produce the desired effect than were required with synthetic chemicals."

There are two routes by which SEB infections can be contracted, and they give rise to different symptoms in human subjects. When inhaled as an aerosol, the toxin can produce fever, chills, headache, muscle pains, and a cough. When taken orally, as in food, the toxin causes nausea, vomiting, and diarrhea—the classic symptoms of food poisoning. Either way, the resulting infection tends to be incapacitating only, not fatal.

The trials were run at dusk, at the time of the day when there was often an atmospheric "inversion," a condition much sought after by biological warfare researchers. Normally, air temperature decreases with altitude, with warmer air below and cooler temperatures above. But the opposite effect can also occur, when a cooler air mass near the ground is overlaid by warmer air. Such a condition is sometimes responsible for producing early-morning or early-evening ground fogs. The dispersal of a biological agent into such a thermal inversion layer has the effect of holding the infectious agent close to the ground, thereby providing maximum "time on target," instead dissipating harmlessly, and therefore uselessly, upward into the clouds.

And so across a series of evenings in the fall of 1968, an F4-E Phantom jet streaked down out of the sky near Eniwetok Atoll and dropped to an altitude low enough for a precise and close dispersal of the agent. The craft approached one of the smaller islets of the atoll, Lowja Island, which the military had code-named "Ursula." A ninety-one-meter tower stood on Ursula Island and recorded atmospheric conditions at the function time. Downwind from the islet was the standard array of five light tugs, caged monkeys, and aerosol sampling devices. Over this collection of boats, animals, and air samplers, the Phantom jet smoothly delivered its infectious cargo.

The ultimate point of these tests, the very last large-scale, open-air sea trials of the Deseret Test Center's biological warfare program, was to establish the potential area coverage and associated casualty levels of this new weapons system, the specific combination of agent and disseminator that was used at Eniwetok. An unclassified paragraph in the final report on the tests summarized their results:

> The weapon system disseminated the aerosol over a 40–50 km [25–30 mile] downwind grid, encompassing a segment of the atoll and an array of five tugs. Stability of the bulk agent and of the agent aerosol was evaluated by the response of animals to the intravenous injection of graded doses. The agent proved to be stable and did not deteriorate during storage, aerosolization, or downwind travel. A single weapon was calculated to have covered 2400 square km, producing 30 percent casualties for a susceptible

population under the test conditions. No insurmountable problems were encountered in production-to-target sequence.

An area of 2,400 square kilometers is equal to 926.5 square miles, more than three times the size of all five boroughs of New York City combined, which encompasses only 302.6 square miles. Assuming a population of eight million New Yorkers and a 30 percent casualty rate, this result means that a single biological weapon could produce approximately 2.4 million casualties—that is, people incapacitated, not deaths. Within the ranks of American biological warfare scientists, the results of Speckled Start, DTC 68-50, only reinforced their longstanding conviction that biological agents and their associated delivery systems constituted a credible, effective, and reliable means of inflicting casualties on an enemy.

By the time that both the POBSP and Project SHAD were over, the US Army had officially standardized, weaponized, and stockpiled two lethal agents, including the tularemia microbe, as well as three incapacitating biological agents, and one incapacitating toxin, SEB. The Pacific Project's fieldworkers were neither directly nor even indirectly responsible for any of this, as the tests were held without their knowledge, consent, or participation—except for the isolated case of Robert Standen, who passively monitored the Baker Island mosquito trials.

In return for its large and sustained investment in equipment, personnel, and money, the army got great amounts of information about birds, their migration patterns, what they ate, what diseases they did and didn't have, and so on. But little of that information was ever put to actual use. It is true that the army had planned a series of additional tests, with unknown agents at unknown locations ("ghost tests"), and had also contemplated tests using the VEE agent, but refrained from doing so in light of experiments at Detrick showing that sooty terns might carry that agent to human populations, and infect them. The VEE tests, and the "ghost tests" were ultimately cancelled, but that made no difference in the final accounting. Indeed, the army could have conducted the same set of tests, with the same results, had the POBSP never existed. And so in retrospect, it must have seemed to the army that its enormous investment, and its seven-year-long relationship to the Smithsonian, was largely a waste of time and money.

The Smithsonian, and ornithological science in general, however, got far more from the connection between the two parties. It may be argued that the Smithsonian acquired this knowledge by accepting tainted funds and that this somehow defiled, corrupted, or otherwise undermined the science. However that may be, the institution's acceptance of the army's money caused it a certain level of embarrassment once news of the program leaked, which was soon enough.

9

The "Secret" Emerges

The Pacific Project was barely even begun before it started appearing in the news, some of the stories implying that the Smithsonian had been duped into or was knowingly complicit in performing covert biological weapons research for the army. The first of these reports charged that the institution's scientists were conducting exploratory studies into the transmission of diseases from birds, animals, and insects to humans, and that they were doing this work not for pure and disinterested research purposes but on behalf of the Department of Defense, which was paying for the whole enterprise. That was the scenario advanced by William E. Small in the journal *Scientific Research*, under the title "DOD Supporting Bird Studies in Pacific, Brazil," in its December 9, 1968, issue.

The piece quoted Sidney Galler, the Smithsonian's assistant secretary for science, acknowledging that the institution was in fact pursuing such work but maintaining that the scientists involved were "free from pressure by the military, [and were] conducting research of their own choosing just as any scientist would under a similar agreement." The story also quoted Philip Humphrey, director of the Pacific Ocean Biological Survey Program (POBSP), as saying, "What they [the DOD] do with the data I don't have any idea. We just send them copies of our results." (And, apparently, keep our eyes closed and hope for the best.)

The news brought forth immediate condemnation from the ranks of academia. The January 6, 1969, issue of the same journal published a letter to the editor by Rockefeller University cell biologist Philip Siekevitz saying that the Department of Defense's use of the Smithsonian was an example of the "adventuristic, imperialistic American military policy, the same as is causing the disaster in Vietnam." He further claimed that it was time for US scientists to "get down to really thinking about our roles in society."

And there, for a short time, the matter rested. Only to be revived a month later, when on February 4, 1969, NBC News ran an episode of its new *First Tuesday* investigative journalism show, in a broadcast anchored by Sander Vanocur and Tom Petit. Petit reasserted the Smithsonian–military connection and charged that the Smithsonian was being used as a "cover" for US Army chemical and biological warfare (CBW) tests in the Pacific. "There has even been an ultra-secret test project in the Pacific Ocean, conducted under cover of a bird-banding study," he said.

Later in the program, Petit introduced Robert Standen as a former POBSP fieldworker. Standen had in fact sailed on Southern Island Cruise No. 8, in May 1965, a voyage that landed him on Baker Island. This was the same month during which the Deseret Test Center was running the Magic Sword mosquito trials there.

Petit claimed that "Standen later took part in an ultra-secret military CBW project in the Pacific." And then, onscreen, Standen disclosed that he had indeed taken part in the test, which he said had involved a "biological carrier." Petit said that the carriers were "animals," and that army, navy, and Air Force personnel were "testing animal vectors, or carriers, to see how they would behave in a tropical climate. No germs were involved. In effect it was a checkout of an animal delivery system for CBW."

All of which was true enough. The carriers in question had in fact been mosquitoes, which were of course members of the animal kingdom. But without saying so in as many words, Petit gave the impression that larger animals, and perhaps even *birds!*, were the vectors being tested—an early, veiled hint of the "bird bomb" theory.

Irrespective of what the carriers had been, a bona fide Smithsonian Institution scientist's participation in an actual biological weapons field trial was blockbuster news that was picked up the next day by print media. The *Washington Post* ran a story under the headline "Smithsonian Bird Research Tied to Germ Warfare Study." The piece quoted Seymour Hersh, author of the book *Chemical and Biological Warfare*, as saying that the Defense Department needed an island free of birds to conduct such tests because otherwise birds could pick up diseases from the tests and then could go on to spread them to human populations elsewhere.

Those claims, however, made little sense. To begin with, if the Defense Department really needed an island that was free of birds, Baker Island would hardly qualify. Between March 1963 and February 1965, the POBSP made seven separate Southern Island Cruises that stopped at Baker. In every instance, the fieldworkers recorded numerous birds of multiple species, including a great profusion of sooty terns. Indeed, the whole notion that there were "bird-free islands" lurking somewhere in the Pacific, or elsewhere, was dubious in the extreme. As we have seen, there were birds even on specks of land as tiny and unwelcoming as La Pérouse and Gardner Pinnacles.

Second, NBC News stated, correctly, that "no germs were involved" in the tests, which meant that no diseases could have been carried away from them by birds, or any other animal, in the first place. But as will become clear, critical reactions to news of the Smithsonian's "secret" were often contradictory, confused, and riddled by half truths and doubletalk.

On the same day of the *Washington Post* story, the *New York Times* ran a news feature claiming that "the Army under the guise of a bird study by the Smithsonian Institution, is looking for a remote Pacific site to conduct experiments in chemical-biological warfare." Suddenly, the Smithsonian was transformed from a benign repository of artifacts and a fount of knowledge into a furtive tool of the army's germ warfare machine.

The Smithsonian, for its part, defended itself as well as it could against these allegations. Its press office issued a statement denying all charges and claiming that the POBSP was "a basic research program consistent with the Institution's traditional scientific pursuits." The project's scientists were collecting data about Pacific Island birds merely "to permit broad ecological conclusions to be drawn." Sidney Galler dismissed the view that the Smithsonian was "an unwitting dupe or cloak for some kind of biological warfare research" and also rejected the notion that the institute was helping the army find a biological weapons test site. But in fact, the POBSP was doing exactly that, whether Galler was aware of it or not.

Galler also said, however, "We have a commitment as scientists and engineers to find the truth. If our nation can make use of that knowledge for national defense purposes, should we be sorry? Knowledge and truth are universal and, so long as we seek them openly and honestly, we never need to be ashamed."

S. Dillon Ripley, secretary of the Smithsonian, was even more glowing about the biological survey program. "It's a wonderful project from the scientific point of view," he said. "The fulfillment of a dream."

Well. Plainly, this mix of charges, rebuttals, and effusions cried out for some sort of an independent assessment, and *Science* magazine soon provided one. In its issue of February 21, 1969, under the title "Biological Warfare: Is the Smithsonian Really a 'Cover'?" author Philip M. Boffey looked at the incriminating claims and evaluated them one by one. First of all, what did it actually mean to say that the Smithsonian was being used as a "cover"?

"In modern spy terminology," he wrote, "the word would seemingly imply either that (i) Smithsonian scientists carried out military activities while pretending to be engaged in research, or (ii) military personnel posed as Smithsonian scientists, or (iii) the Army, in order to hide its intentions, used the Smithsonian to perform research that should normally have been performed by the Army itself. None of these seems to have been the case."

But in fact, version (iii) came pretty close to the actual truth. Clearly, the army was withholding full disclosure of its true motivations from the Smithsonian. But at the same time, the army was in no position to do ornithological survey work itself and for that reason commissioned the Smithsonian to do it for them. The army was *using* the Smithsonian for its own purposes, whether or not as a "cover."

The *Science* piece quoted Philip Humphrey as saying that "the project is not Army-directed research—it's Smithsonian research supported by the Army." But in fact it was both. The army not only initiated the program, but it also dictated the size and scope of the survey, and established the range and location of the islands and areas of the open ocean to be visited, required the capture and shipment of hundreds of live birds as well as the collection of blood and spleen samples, and so on. Without question, that was "Army-directed research."

Humphrey made a better case against the further claim that the army was specifically developing a bird delivery system for germ warfare, a notion that he dismissed as "ridiculous": "While birds in a statistical sense may have predictable migrations, in an individual sense you don't know what the hell they're going to do." That point was well supported by the project's bird-banding returns: although clearly defined migration patterns existed, individual birds were recaptured at random and unpredictable points all over the Pacific. Other than homing pigeons, birds are unreliable couriers.

Finally, Boffey contacted Robert Standen, whom he called "NBC's star witness," and asked him about the precise nature and extent of his participation in the Baker Island biological weapons tests. "Standen said that the Army asked the Smithsonian project to send an observer along so that, if the test caused biological changes on the island, the Smithsonian scientists would understand what had happened. As it turns out, Standen said, there were no changes, so Standen left the island after 12 days, well before the end of the test." In other words, he was an innocent bystander playing an essentially passive role.

In light of these facts, Boffey concluded that "NBC's use of the word 'cover' to describe this situation seems highly misleading." As indeed it was. And so the net effect of the *Science* story was to substantially defuse the image of the Smithsonian as an agency knowingly or unwittingly engaged in biological warfare operations.

Not everyone was satisfied with that apparent exoneration, however. In a letter to the editor, Harvard University evolutionary biologist Stephen Jay Gould argued that the Smithsonian's acceptance of tainted army dollars to further its own research agenda amounted to a Faustian bargain. "Our professionalism has distorted the ranking of our values," he said. "We have so inflated the importance of our research that we silently accept heavy strings on doubtful money to pursue work that would otherwise not be funded. We commit, in other words, the classic sin of pride."

That was another distortion of the truth. The Smithsonian did not "accept heavy strings" in return for the army's money. The only constraint the institution accepted was an obligation to submit articles intended for publication to the army in advance for prior clearance. And there is no available evidence that the army ever exercised its prerogative to withhold or censor any publications by the institute's researchers. In practical effect, therefore, the army's support was a "no-strings-attached" grant.

Among the parties most disturbed by the Pacific Project's covert army funding was the Smithsonian's own Senate of Scientists. This was a body, established in 1963 by the scientific research staff of the National Museum of Natural History (NMNH) "to address their professional concerns," a Smithsonian statement said. "Modeled on faculty senates in universities, the Senate was structured to function as a source of collective opinion outside normal administrative channels." That is, its role seems to have been advisory rather than legislative or binding.

Over the course of its existence, the senate addressed "issues of Smithsonian governance, library service, parking policies, off-mall storage and curatorial facilities," and the like. In addition, it aimed to promote "collegiality within NMNH, through its publications, seminars, teas, and dinner forums."

In a memorandum dated November 17, 1969, the senate formally expressed its disapproval of the Pacific Project and called for its termination "without reservations of any sort on June 30, 1970. Simultaneous termination of Philip Humphrey as Research Associate in the Department of Vertebrate Zoology would also be viewed with favor by the NMNH Senate members."

In point of fact, Humphrey, who had been chairman of the institution's Department of Vertebrate Zoology, left the NMNH in 1967 to become the director of the Natural History Museum at the University of Kansas, although he was nevertheless kept on at the Smithsonian as a research associate. But the Senate of Scientists wanted him all the way banished and gone, in effect to become a nonperson.

And soon enough he was. Whether the NMNH senate's recommendations had any official clout or force, the POBSP was indeed terminated by June 30, 1970. And by the end of the year Philip Humphrey was no longer associated with the Smithsonian Institution in any capacity.

The upshot of these events was that everyone involved with the Pacific Project was temporarily under a cloud of having played a role in something that was at best controversial and at worst slightly improper.

"The stigma of working on the Pacific Project was never fully erased during my time with the Smithsonian," Fred Sibley remembered. "Although many people in hindsight said, 'I wish I had joined.'"

Indeed, the only party relatively untainted by the whole affair was the US Army itself. It never pretended that its biological warfare work was anything but its own secret business.

* * *

In none of the early news accounts of the Pacific Project's relationship to the army's classified biological warfare program was any mention made of the Smithsonian's prior work for and cooperation with the US military, which was considerable. Some of the Smithsonian's earlier efforts on the military's behalf even involved classified projects, a fact that had bothered nobody at the time.

Much of this work had occurred during World War II, when the idea of cooperating with the army, navy, air force, and marine corps was viewed in a wholly different light than it was in the 1960s, during the Cold War era, when the POBSP took place. In wartime, Smithsonian administrators, curators, and scientists believed they had a patriotic responsibility to do what they could to help win the war.

The bombing of Pearl Harbor turned everyone's eyes toward the Pacific, an area about which Smithsonian scientists had already acquired a substantial body of expert knowledge. Different specialists were familiar with the botany, zoology, customs, languages, geography, and natural resources of the region. The scientists were also skilled researchers who knew how to access information that was not otherwise easily available in an age well before the dawn of the internet, Google, and Wikipedia. In her scholarly paper, "The Smithsonian Goes to War: The Increase and Diffusion of Scientific Knowledge in the Pacific," Pamela M. Henson, a historian at the Smithsonian's Institutional History Division, described some of the ways the institution's scientists aided the military before, during, and after World War II. Because the Smithsonian was a seemingly inexhaustible source of knowledge, one of its roles after the start of the war was to fulfill informational requests from members of the armed forces about a wide diversity of subjects.

"In the first year of the war, the Smithsonian received over 700 such requests," Henson wrote. Military planners asked for "information from physical anthropologists on the range of normal human head shapes for designing gas masks; a study of Eskimo clothing in the collections of the National Museum to help design Arctic gear; methods of securing drinking water from seawater; locations of new sources of supply of quinine, fibers, and rubber from botanists in the National Herbarium; and, last but not least, maps of the migration patterns of snapping shrimp, so ships could travel with them and confuse Japanese sonar."

Another request concerned a little-known Japanese attack on the Aleutians. Starting in June 1942, in the only actual occupations of the United States by a foreign power during the war, a small Japanese force took hold of the two most remote Aleutian Islands, Attu and Kiska. Attu is the westernmost island of the Aleutian chain, and therefore of the United States. Indeed, it is so very far west that the island is not even in the Western Hemisphere. Attu lies on the other side of the 180° meridian, placing that part of the country in the *Eastern* Hemisphere and far closer to Russia than to mainland Alaska.

At the time of the Japanese invasions of Attu and Kiska, the US military knew virtually nothing about either place. But as Henson writes, "Henry Bascom Collins, an anthropologist with the Bureau of American Ethnology who had conducted archeological fieldwork in the Aleutians, immediately turned over his maps and photographs of the islands and provided information about languages, customs, and locations of settlements, as well as appropriate sites for airfields and military bases."

After a year of fighting the Battle of Attu, the United States proclaimed victory over the Japanese invaders and reclaimed the two islands on August 15, 1943.

Tales of downed fliers on isolated Pacific islands and atolls prompted Smithsonian researchers to fulfill a navy request to produce a handbook for marooned personnel. The result was a 187-page waterproof pocketbook, *Survival on Land and Sea*, a compendium of practical, life-saving information and survival techniques. Almost a million copies of the book were printed and distributed before the end of the war.

The Smithsonian provided further assistance to the military even in the postwar era. In 1946, the navy asked the institution for expert assistance in collecting data about wildlife conditions at Bikini Atoll, both before and immediately after the two atomic bomb tests conducted there. While the Bikini tests were not secret, parts of the operation were classified, but this seems to have caused no uproar at the Smithsonian. And so the institution duly sent the navy two seasoned field collectors who essentially ran their own mini-POBSP-style biological survey of the place, just before and after the detonations. Unfortunately, their results were lost in a shipwreck off the California coast on the return trip.

The same two Smithsonian scientists revisited Bikini Atoll a year later, but according to Henson they "did not think that much could be concluded within a year of the blast. Many organisms had died from the impact or from radiation poisoning. But little mutation was seen. They speculated that competition on the reef was so keen that any aberrations were gobbled up long before they could be collected by scientists. . . . Plankton glowed on photographic plates, as did the intestinal tracts of the fish that fed on them. Only long term studies would show if the atoll would ever return to the ecological balance it enjoyed before Able Day [the date of the first atom bomb test]."

Later, other scientists would come to the islands and attempt to restore their disturbed ecologies and their physical conditions to their prior and more natural states. Such activities were no part of the Pacific Project itself, at least not officially. On Baker Island, however, a few Smithsonian fieldworkers—Fred Sibley, Binion Amerson, Larry Huber, and Roger Clapp among them—made determined, and for a while apparently successful, attempts at ridding the island of its cats in order to let bird populations there recover and flourish. But with the temporary disappearance of the cats, the island's mouse population exploded, and in July 1964 mice were described as "abundant" on Baker. Thereby illustrating once again the difficulty of optimizing all aspects of a given island's ecological balance simultaneously.

* * *

The army's final bioweapons tests in the Pacific, series DTC 68-50, ended in October 1968. POBSP field trips continued through 1968, however, with the last of the Southern Island Cruises, No. 20, running between October 7 and November 3 of that year. The army extended the program through June 1970 to give the researchers time to write up their findings.

The army's ultimate motivation in funding the Pacific Project was to advance biological weapons technology, theory, and practice. But in a major surprise move a year after the final POBSP Southern Island Cruise, President Richard M. Nixon formally terminated the American biological warfare program altogether. On November 25, 1969, in a press conference held in the Roosevelt Room of the White House, and in a formal "Statement on Chemical and Biological Defense Policies and Programs," Nixon said, "I have decided that the United States of America will renounce the use of any form of deadly biological weapons that either kill or incapacitate."

Further, he ordered the Department of Defense to begin the process of destroying existing stocks of bacteriological, viral, and rickettsial weapons. All future research would be confined exclusively to defensive measures against biological attack. Those measures included the development of vaccines, antibiotic treatments, and other methods of controlling and preventing the spread of bacterial, viral, and rickettsial diseases that might be used against the United States by other nations.

Nixon's action meant that as far as the army was concerned, all of the POBSP's findings were suddenly of no interest, relevance, or value. Overnight, the "military payoff" of the Pacific Project went up in smoke.

For members of the US Army Chemical Corps, the cancellation of its offensive biological warfare project was hard to take. The obvious question at the time was: *Why this sudden change?* To this, Nixon's *professed* answer was that "biological weapons have massive, unpredictable, and potentially uncontrollable consequences. They may produce global epidemics and impair the health of future generations. Mankind already carries in its hands too many of the seeds of its own destruction. By the examples that we set today, we hope to contribute to an atmosphere of peace and understanding between all nations."

To old-line biological warriors, this was not the real answer. In their view, biological weapons were not inherently "unpredictable." To the contrary, the scientists had amassed large datasets characterizing the behavior of biological agents in a wide variety of environments and atmospheric conditions. They could recite chapter and verse about agent decay rates, downwind travel and dispersal patterns, the concentration of viable pathogen per unit of elapsed time, the total number of likely casualties in various population settings, and so on.

Moreover, Nixon's claim that biological weapons could "produce global epidemics" was flatly untrue, at least insofar as it applied to United States' own storehouse of germ weapons. The only biological agents the nation had ever weaponized and stockpiled were those that had low or nonexistent rates of person-to-person transmission and secondary spread. Anthrax, for example, one of the army's first and chief biological weapons, did not travel from person to person. In nature, anthrax outbreaks arise from sources such as contact with infected animals or contaminated animal products.

Nixon's inaccurate characterizations of the predictability and transmissibility of the army's biological weapons led to speculation as to what were his actual motives in banning them. After all, the sudden ban flew in the face of the US military's longstanding rationale for creating, testing, and stockpiling biological weapons; namely, their deterrent value. The argument had always been that other nations would be deterred from using of such weapons themselves if they knew that the United States maintained a massive store to be used as an in-kind response to a biological attack. If that deterrent value were to fail and the nation came under a biological first strike, then the army had a retaliatory biological weapons stockpile readily available for an in-kind response. Outlawing all biological weapons would remove both of those protections in one fell swoop.

In fact, the actual motives behind Nixon's decision were quite different from the reasons he gave publicly. In a sort of replication of the Kennedy administration's military preparedness review, the incoming Nixon administration also took a fresh look at the nation's weapons systems. In this case, however, the assessment was instigated not by the president or his secretary of state but by Congress.

The basic sequence of events was that soon after Nixon took office in January 1969, members of Congress asked the president to clarify the nation's policy regarding biological and chemical weapons. The request was prompted by controversies associated with three recent incidents concerning chemical (but not biological) munitions. The first was an accidental release of the nerve agent VX during an open-air test at Dugway Proving Ground in March 1968. The lethal chemical killed more than six thousand sheep.

Second was the revelation during the spring of 1969 that the army was transporting leaking chemical munitions across the country by train and loading them onto ships, which were then scuttled at sea, in a project known as Operation CHASE (cut holes and sink 'em). Transport of the chemicals over land posed obvious risks to neighboring populations, and ocean dumping presented a range of potential adverse consequences for marine life.

Third was the international condemnation of the US Army's use of the chemical defoliant Agent Orange in Vietnam. Agent Orange use was also opposed by a significant proportion of the US public.

In light of these pressures, Nixon ordered a reassessment of the utility of chemical *and* biological weapons. On May 28, 1968, National Security Advisor Henry A. Kissinger issued a Secret (later declassified) "National Security Study Memorandum 59," which stated in part:

> The President has directed a study of U.S. policy, programs and operational concepts with regards to both chemical and biological warfare and agents.
>
> The study should examine present U.S. policy and programs on CBW, the main issues confronting that policy, and the range of possible alternatives thereto.

In its way, this was a reprise of Robert McNamara's Project 112 eight years earlier, although it authorized no testing, only an exhaustive analysis of the pros and cons of biological and chemical munitions. A central issue was weighing the benefits of maintaining a germ warfare capability against in the face of its potential costs and drawbacks, of which, it turned out, there were many.

For one, biological agents had a limited lifetime in storage and had to be replenished annually. For another, they were slower acting than chemical munitions, conventional explosives, and atomic weapons, and were less reliable in the field. The microbes had an incubation period of several days to a week or more, which meant that the enemy could be exposed to a pathogen and not be aware of it, or experience its ill effects, for days at a time. For that reason, biological weapons had little utility in actual battle. They were better suited for mass attacks on population centers, but that is an indiscriminate use that would cause illness and deaths among innocent noncombatants, which is morally repugnant. Finally there was the fact that some microbial pathogens were persistent agents and might do damage long after a given conflict ended. They might even mutate into something worse than they were to begin with.

For these reasons, on August 15, 1969, Defense Secretary Melvin Laird ordered the temporary suspension of biological agent production. Politically speaking biological warfare had become a what he called a "tar baby," a difficult problem that is only aggravated by attempts to solve it. By November of that year, Laird came to an even more negative conclusion. "Biological research is something that can be supported, but biological warfare cannot be supported by anyone."

Separately, Nixon had both political and public image reasons for renouncing biological weapons. Abolishing them, he thought, would send a message to other nations that such weapons were not worth having, which would discourage hostile states from acquiring them, or maintaining their stockpiles if they already had them. Additionally, the act of giving up bioweapons had immense public relations value, as they were widely considered morally indefensible. In banning them, Nixon could come across as statesmanlike and a man of sanity, probity, and peace. On balance, there was more to gain by renouncing biological weapons than there was by retaining them. Therefore, they had to go.

This was a significant move on Nixon's part because it was the first time that a major world power unilaterally relinquished an entire category of armament. But, as fate would have it, Nixon's abandonment of biological weapons failed to achieve one of its primary objectives: it did not cause other nations to do likewise; in fact, just the opposite. The Soviet Union, in particular, decided that Nixon's renunciation was merely a hoax and the United States was secretly maintaining its biological weapons program, but hiding it well out of public view. In turn, the Soviets secretly kept up, and even enlarged, the size and scope of their own bioweapons research, development, and biological agent stockpiles.

As for what Nixon's decision meant for the POBSP, it made the entire survey program into a militarily gratuitous enterprise. Its utility to the army, low enough to begin with, now simply vanished altogether. All of the project's research findings, its meticulously cataloged records of species observations, bird counts, bandings and recaptures, blood samples—everything that the POBSP had learned about species populations and distributions across the Pacific—all of it instantly lost its military significance. It was as if the entire project had never existed.

* * *

The project's scientific value was quite another matter. The sum total of the Pacific Ocean Biological Survey Program's activities far transcended its military raison d'être. Its scientific contribution must be evaluated on its own terms, separately from the narrow and utilitarian purposes envisioned by the army.

The Pacific Project was a baseline survey: it provided a record of what was actually out there in the world, a snapshot of the status quo within its domain at or across a given moment in time. That information will be useful to future researchers assessing how and to what degree those conditions have changed in the interim.

In 1965, toward the beginning of the project, Philip Humphrey wrote what was and still remains the best summary of the program's value to science:

> Perhaps the most important practical accomplishment of the Smithsonian survey will be the delineation of the environment over a relatively short period of time. This will provide a baseline of comparison for biologists concerned, 10 or 20 years from now, with measuring the effects of man-made modifications of the environment on natural populations of organisms. The need for such a baseline is most urgent today, when man, in his struggle to advance himself, is changing the face of the earth at an appallingly rapid rate, and is subjecting the total environment—water, atmosphere, and living tissues—to physical and chemical influences which need to be measured now and in the future. For unless these fundamental changes in his environment are properly assessed, man himself, through ignorance, may fall victim to his own progress.

But baseline studies have their limitations, and this was no exception. Much of the information gathered, especially the epic amounts of bird bandings, amounted to vast masses of data, with large quantities of it remaining unanalyzed and unutilized even to this day. The amount of data generated and recorded by the project's scientists is overwhelming in its sheer bulk: 32.1 linear meters (105 linear feet) of paper documents, photographs, maps, and assorted other items, stored in 245 boxes in the Smithsonian Institution Archives. The archive's finding aid, or catalog, of this collection itself runs to 119 pages, the final entry of which, in a file named "Added Accession," is designated "Last Minute Things."

Over and above being a cache of files and facts, the survey also produced tangible, physical results: thousands of bird skins, some of which are on public display at the National Museum of Natural History, others of which lie in drawer upon drawer in the vast Smithsonian bird collection repository. Those collections constitute an important reference tool for a researcher looking to document a possible new bird species or to verify a known one. Altogether, they amount to a library of birds.

The Pacific Project failed to produce any new theory, large or small, and yielded few major or minor insights into or about biology or evolution, or other fundamental questions regarding the workings of nature. As Roger Clapp once joked, "We didn't do Darwin, Darwin did Darwin." But in an obvious way, much of what the fieldworkers saw and documented was "new," if only in the rather trivial sense that it had been previously unknown or unnoticed and unrecorded by scientists, or, in many cases, by anyone else.

In addition, the survey was responsible for several discrete discoveries that, although they may be minor in themselves, were nevertheless surprising to specialists. For example, the fact that ruddy turnstones fly from Alaska to Hawaii; the discovery of at least two new species of chiggers, which were named after Binion Amerson; the finding that northern fur seals had re-established a breeding colony in California after being absent from the area for more than one hundred years.

The program also notched a number of firsts. For example, Pacific Project's scientists were the first to employ computer processing (using the then-current medium of punch cards) to analyze "big data" in ornithology. The Smithsonian fieldworkers were also some of the first scientists to experiment with radar as a means of tracking seabirds, a methodology that did not come into widespread use until much later, in the 1990s. In 1965 and 1966, on St. George Island in the Aleutians, Max Thompson and Robert De Long pioneered the use of cannon-projected and rocket-propelled nets to capture and band shorebirds, sometimes netting as many as 175 ruddy turnstones at a time.

But beyond the program's individual achievements, discoveries, and firsts, one of the POBSP's greatest contributions to science was in furthering what we might describe as "the education of our heroes." After all, the fieldworkers included a bunch of kids fresh out of college who were suddenly thrust into the middle of the Pacific, where they saw and surveyed island after island and inventoried their life forms, in a manner that few have done so fully and systematically, before or since. For them, that was an education in the truest, highest, and best sense of the word.

And that education did not end with the termination of the program. For although most of the project's personnel are today dead and gone, the products of their work survive in the form of written records, reports, and other documents. These items run the gamut, ranging from scribbled field notes—many available in digital form on the Biodiversity Heritage Library website (biodiversitylibrary.org)—to unpublished manuscripts available at the Smithsonian Institution Archives (Record Unit 245), to journal articles large and small, to detailed book-length accounts published in the Smithsonian's *Atoll Research Bulletin*. Many of the *ARB* reports are online and downloadable as fully searchable PDF files. In total, the POBSP authors were responsible for producing more than 175 individual accounts of the natural history of the Pacific Ocean and its islands and atolls, their bird species, and other wildlife.

The *Atoll Research Bulletin* monographs are particularly valuable for having rescued from otherwise probable oblivion the stories of when, how, and by whom even the smallest specks of land were discovered, the use that different visitors made of them—explorers, colonizers, whalers, guano miners, wrecked sailors, the armies and navies of various nations—and their eventual fates, at least to the time of publication. The preservation of that history is perhaps the Pacific Project's most lasting contribution to science, one that has long since effaced the stigma of its origins from within the depths of the US biological warfare program.

10

Fate of the Islands

The Pacific Project was a multifaceted undertaking that encompassed Smithsonian scientists, administrators, fieldworkers, and supporting staff. The project also embraced the enlisted personnel and officers of the army and navy who transported the field teams from place to place all around the Pacific. It included officers of the four military services that made up the Deseret Test Center and gave the fieldworkers their marching orders. And by extension, it even included the scientists at Fort Detrick who made specific bird, tissue, and other biological sample requests to the field teams.

But the project could not have existed at all were it not for a further set of participants that constituted the fundamental locus of operations, which is to say, the various islands of the Pacific, together with their bird, mammal, and plant populations. These islands were affected by the Pacific Ocean Biological Survey Program (POBSP) in at least two ways. First, negatively, as the field team members killed and skinned birds for specimens. They captured other birds live, crated them, and shipped them away—sometimes so many that they made a noticeable dent in species population levels. Project scientist Fred Sibley said, "At one point a number of us signed a letter questioning the number of birds being removed as some of the species populations were small."

But in another way the islands were affected positively as field team members deliberately worked to protect the birds. For example, the men made repeated attempts to eradicate invasive species such as cats that preyed upon birds to the point that they sometimes eliminated entire species from their breeding grounds.

A few islands, such as Birnie, experienced little change in the years following the end of the survey project, while others had their ecologies altered significantly, and often damaged. Some places continued to be visited by former Pacific Project fieldworkers taking part in similar biological study programs sponsored by other agencies afterward. Still other islands would go on to have their disrupted ecologies restored by people— volunteer conservationists, ornithologists, US Fish and Wildlife Service personnel—who had no connection to the POBSP at all. And one of the locations, Johnston Island, the centerpiece of the artificial atoll, would go on to suffer a particularly bizarre turn of fate.

By far the greatest damage done to the islands and their bird life was caused by the US military. In 1972, Warren B. King, a POBSP scientist from the beginning of the program, published an overall summary report, *Conservation Status of Birds of Central Pacific Islands*, focusing on the islands visited by the Pacific Project field teams. "The report will stress man's influence on the islands," he wrote, "even though in some instances it would be difficult to show direct causal relationships between man's activities and deterioration of the bird fauna. In other cases it is all too blatant."

One of the most blatant of the cases was that of Sand Island, part of Midway Atoll. The island had been the site of a US naval air station since 1941, and by the 1960s it was home to some two thousand military personnel. Although the navy maintained an official policy of not harassing the birds, which was enforced by a fifty-dollar fine for the killing of an albatross (popularly known as "gooney birds"), construction of the island's three runways necessitated "the destruction by asphyxiation of 18,000 incubating albatrosses, 13 percent of Midway's and about 5 percent of the world's population," King wrote. "Recently all active nests on Sand Island's golf course have been destroyed to keep the area open for recreation. Here we have an example of the ethical paradox by which government sanctioned mass killing is permissible, while the same activity conducted on a small scale by individuals is heavily penalized."

Midway's other land mass, Eastern Island, held large populations of frigate birds and boobies and a substantial colony of sooty terns. "Unofficial policy, confided several times to POBSP personnel engaged in fieldwork on Eastern Island, was the encouragement of the destruction of Sooty Tern eggs, young and adults. Navy personnel participated in 'chick-stamps,' and admitted to clubbing adult Sooty Terns from the air with sticks on several occasions. Dogs were brought from Sand Island to Eastern for the express purpose of running them through the incubating albatrosses."

Other such military atrocities also occurred elsewhere in the Pacific, generally on a lesser scale of callousness. In some instances, though, the army acted to improve the environmental state of an island.

* * *

Two of the places most frequently visited by the POBSP were the twin equatorial islands, Howland and Baker. Between the years 1963 and 1967, POBSP fieldworkers made twenty-five trips to Howland Island and twenty-three to Baker, for a combined total of forty-eight separate visits. By any standard that was an extraordinary number of voyages to islands each of whose land areas totaled less than a square mile and whose ecologies did not change much from one visit to the next.

The fieldworkers themselves could not adequately explain the frequency of those visitations. "I haven't a clue why we did so many to those two islands except they were the first islands after leaving Johnston," said Max Thompson, who himself made multiple stops at both places.

Adding to the puzzle was the fact that no official POBSP publication was ever issued for either Howland or Baker—neither in the *Atoll Research Bulletin*, nor in another scientific journal, nor anywhere else. Expedition crew members produced a few unpublished preliminary reports on the twin islands, and these were presumably sent to the army. The POBSP made far fewer voyages to the third island in the HBJ group, Jarvis, stopping there only six times across the years from 1964 to 1968. And likewise, only unpublished draft reports, field notes, and data books exist for Jarvis Island.

Howland, Baker, and Jarvis share many common features. All three lie in the Equatorial Dry Zone and are so exceptionally reflective of sunlight that they can actually repel rainfall by means of convective heating: it can be raining in the surrounding ocean waters, while the islands themselves remain dry, the rain evaporating before reaching the ground. They also share a common history and geography, all of them remnants of ancient volcanoes that rose from the seabed. All of the islands suffered similar human disruptions in the form of use by whalers, guano miners, and (except for Jarvis) military installations, assaults, and defenses. Human occupation brought rats from various ships and shipwrecks. Later, the guano miners brought cats to the island for rodent control, as did the Hawaiian colonizers. Both the miners and the colonizers also bought in dogs as pets, as well as other invasive animal and plant species, for a variety of purposes. All these uses and foreign species introductions had substantial effects upon the ecology of the islands, causing losses in their bird populations and in species diversity.

The three islands were already US possessions by the time of the Pacific Project surveys, and later, in 1974, all three became parts of the US National Wildlife Refuge. But for all their common features, each charted its own separate path toward a return to a state approaching "normalcy," which is to say, an ecological state resembling conditions beforehand, prior to human occupation and disruption. Meaning, essentially, the status quo before the arrival of humans, rats, and cats.

Baker Island was the first to be rid of all three species. The Hawaiian colonists left the island in 1942, after a series of Japanese air raids and submarine attacks. Rats had disappeared from Baker long before that, by 1937, apparently exterminated by the cats, which continued to thrive on birds, fish, mice, and insects. The cats, in turn, were driven to a temporary extinction by POBSP fieldworkers.

"A total of 27 cats has been killed on the island during POBSP trips," Fred Sibley wrote in his preliminary report on Baker. By July 1964, the POBSP observers saw no more cats . . . until one inexplicably reappeared in May 1965. The last cat was gone for good by 1970. Following their removal, species such as sooty terns, great and lesser frigate birds, and other species resumed their former nesting activities on the island.

Howland was the second of the three islands to be rid of its people, rodents, and felines. The Hawaiian colonists were largely responsible for the rat eradication. "The colonists, like all previous inhabitants, tried to eliminate the rats," Sibley and Clapp wrote in a preliminary report on Howland. "They used traps, clubs, poison and eventually introduced cats in 1937 or 1938. These efforts were evidently highly successful. . . . Sometime after 1938 the rats were eliminated. No rats have been seen on any POBSP trip."

The cats themselves then became the targets for eradication, and a few POBSP members hunted them down almost recreationally. "Starting in March 1963 cats were hunted casually whenever a visit was made to the island and the last cat was killed in February 1964," Sibley and Clapp wrote in 1965. "A total of 26 cats was killed by POBSP personnel."

But their disappearance from Howland was again only temporary; cats showed up again in 1966, in the wake of US military visits to Howland and Baker. After that, the only birds nesting on Howland were those that could withstand cat predation: species such as red-tailed tropicbirds, which could defend themselves; frigate birds, which nested above the ground in kou trees; and masked boobies, whose large size enabled them to ward off cat attacks.

The Howland Island cat population grew slowly but steadily until March 26, 1986, when two wildlife refuge scientists arrived on the island on a quest to kill however many they could find, which they thought was probably about fewer than twenty. The two men were Stephen Berendzen, of the Hawaiian/Pacific Islands National Wildlife Refuge, and Douglas Forsell, of the Alaska Office of Fish and Wildlife Research.

The men brought with them three types of weapons: traps, shotguns, and carbon monoxide cartridges. The purpose of the cartridges was to flush the cat out of their dens so the animals could be then shot. However, they could also be used to cause death by asphyxiation. Armed with this triad of deadly weaponry, it shouldn't have taken them very long to find and kill a mere twenty cats on a flat, treeless island with an area of six-tenths of a square mile. But the two spent seventeen days on Howland and utterly failed to dispose of its cats.

The cat traps proved useless because the island's hermit crabs, which were abundant, crawled into them and ate the bait before the cats did. "These attractants lured hundreds of crabs to the trap site every night," the men reported in their account of the extermination project.

As for the CO cartridges, they at least flushed four cats from their hiding places, which in one case was a wrecked B-26 airplane wing. The men also drove one cat out of its refuge in a patch of dead crabgrass and shot it. Only a single cat died from asphyxiation.

"Shooting was by far the most effective method of killing cats," the men reported. "We frequently saw cats in the interior area of Howland Island and had some difficulty in shooting the last few cats. . . . Some cats may have become very cautious, and we may have missed them."

After seventeen days on the island, the men had killed seventeen cats, for an average kill rate of one per day.

"It was a challenge to remove all the cats," Steve Berendzen remembered much later, in 2020. "I timed our trip to be at the theoretical low point of their population, shortly after nesting sea birds arrived. We spent every night searching with spotlights and several days scouring the island for potential dens and removed a few from them. When we left, I thought there was likely one elusive cat still on the island. My understanding is that a return trip to the island about five years later resulted in finding one old male that they removed."

And with that finally accomplished, bird species that were formerly common returned, and by 2007 there were 150,000 sooty terns on Howland Island.

* * *

That left Jarvis. Jarvis Island was larger than Howland and Baker combined, about one and a half square miles in area. It was a little more heavily vegetated as well, but still treeless, and it was thought to harbor a population of more than a hundred feral cats. With its large area and more numerous hiding places, getting rid of every last one of them would be even more challenging than it was on Howland. In 1964 and 1965, POBSP field teams killed more than two hundred cats, but later visits in 1967 and 1968 showed that there were still eight or nine on the island, more than enough to start repopulating the place, with the potential consequence of further bird mass slaughter.

What would prove to be a prolonged and fitful war against feral cats on Jarvis Island began on June 14, 1982, waged by David Woodside of the US Fish and Wildlife Service, and Mark J. Rauzon, who at that point was a graduate student in geography at the University of Hawai'i at Mānoa. The men arrived by boat from Christmas Island, two hundred miles away, to which they had flown from Honolulu. Because they knew that this would be a protracted and determined eradication campaign, they bought along enough food and water to last them a month.

"Like the stereotypical cartoon, we were two men on a desert island," Rauzon said later. "Only without the palm tree."

There was in fact no shade at all on Jarvis other than what the two men provided for themselves by draping a tarp from the top of a Fish and Wildlife No Trespassing sign. The warning on its front was written in several languages.

For their extermination campaign the men had equipped themselves with *four* different types of weapons: poison, traps, firearms, and a virus. This was the feline panleukopenia virus (FPLV), with which the men were going to wage their own private biological warfare attack upon the cats.

"We were not cat-haters, we were bird lovers," Rauzon explained. "I estimated the Jarvis population of about 115 cats could eat approximately twenty-five thousand birds a year and had already exterminated six species of seabirds from the island."

Using viruses to exterminate invasive species was not a new idea. The method had first been used in Australia, where an initial importation of a dozen or so domestic rabbits for hunting purposes in 1859 led to an explosion of the rabbit population there, until there were many hundreds of millions spread across the continent. The myxoma virus was known to be fatal to domestic rabbits, and in 1950, scientists released it into the Australian Outback. This proved a flat-out success and caused the deaths of an estimated five hundred million rabbits.

However, the same method of biological control was quite ineffective on Jarvis. Rauzon and Woodside fitted five cats with radio transmitters to track their movements, and gave each an oral dose of FPLV. They then dosed an additional twenty-six cats by inoculation, marked them with spray paint, and released them back into the wild. This was supposed to be a more "humane" and "natural" way of killing the cats, but it just didn't work.

Although the feline panleukopenia virus was known to be highly contagious and was often fatal, it did not turn out to be very deadly on Jarvis Island. Only one of the five orally dosed, radio-tracked cats died within ten days. The remaining four were still alive eighteen days later, after which the men simply killed them manually. Of the inoculated, spray-painted cats, ten were still alive and healthy after eighteen days, and wound up being shot.

Poison, trapping, and particularly hunting were far more effective killing techniques, at least in the early stages. "The obvious trend is initially high mortality with a quick drop-off as hunting progresses," Rauzon said. "The total yield for 110 man-hours of hunting was 105 cats or about one cat per hour of hunting."

At the end of their time on Jarvis, with their own radio transmitter dead, their food rations running out, and both men losing weight, they still had not killed all the cats. They left the island anyway, knowing that the birds were at least saved from further feline predation.

That October, three months after their departure, David Woodside returned alone to Jarvis Island. He spent one hundred hours hunting but shot only two cats. He returned again five months later, in March 1983, and spent another hundred hours hunting cats without killing a single one, although he did spot a lonely cat on the island. One cat, however, was enough to cause disaster because a single pregnant female was all that was needed to repopulate the island one more time.

In April 1990, eight years after his first visit, Mark Rauzon came back to Jarvis, determined to finish the business once and for all. He walked the island at night with a flashlight, until, finally, there on the ground in front of him was a pair of gleaming blue eyes.

"I could not leave Jarvis without killing this cat," he said later. "Eradication requires total commitment to completion."

He got off a clear shot and killed the last cat of Jarvis Island.

Rauzon returned to the island on one final occasion, in 1996, which was now fourteen years after the initial eradication attempt. And on this visit he saw no cats. Later, in his 2016 book, *Islands of Amnesia: The History, Geography, and Restoration of America's Forgotten Pacific Islands*, Rauzon summed up the effects of his and Woodside's work over the course of their eight-year-long island restoration campaign.

"When we were there in 1982, eight seabird species bred on Jarvis," he wrote. "By 1996, five more species had recolonized. By 2004, fourteen species were breeding, including the rare Polynesian storm petrel, making its return after some forty years absence. Brown noddies multiplied from just a few in 1982 to almost ten thousand. Grey-backed terns numbered only twelve in 1982; by 2004, about two hundred nests were found. Christmas shearwaters, red-footed boobies, brown boobies, red-tailed tropicbirds, great and lesser frigatebirds, and sooty terns in uncountable numbers have also recovered in the intervening years, making Jarvis the largest and most diverse seabird colony in the Central Pacific. . . . We rescued Jarvis at its nadir, and it's completely recovered from the feral catastrophe."

* * *

The environmental restorations of Howland, Baker, and Jarvis were three clear success stories. Somewhat different fates awaited the two other locations the POBSP had visited and that the US Army and Project SHAD ships, planes, and personnel later used in conducting biological warfare trials: Eniwetok Atoll and Johnson Island.

Prior to the POBSP, between the years 1948 and 1958, the United States detonated forty-three atom bombs on various islands of the atoll. The residual radioactive fallout from these blasts made Eniwetok Atoll a veritable hot zone of the Pacific, hazardous to humans, mammals, birds, and even fish. But island restorations were not done only for the sake of birds and other wildlife. In this case they were also done for the Native populations who had once lived on the atoll and from which they were evacuated before the start of the nuclear weapons testing.

The restoration of Eniwetok Atoll was a major, massive, expensive, and largely unsung part of the US nuclear weapons program in the Marshall Islands. During the years 1977–80, after the completion of a radiological assessment of each of the atoll's islands, the US military undertook a $100 million cleanup project. The effort required more than 6,600 American personnel from three branches of service, plus 1,011 more from the Department of Energy, as well as civilian contractors, journalists, and miscellaneous others. Six US servicemen would die during the project, which, in the end, could not be pronounced an unequivocal success.

The most heavily contaminated soils and debris from forty-two islands of the atoll were transported to another of them, Runit Island, where they were buried in a crater that itself was a product of one of the atomic bomb blasts, code-named "Cactus," in 1958. The radioactive materials were mixed with Portland cement that would harden in place so as to form an immense and immovable block of material when finished. Then, on top of it all, a thick concrete cover was erected. The final result was the 350-foot-wide Cactus Crater Containment Structure, also known as Runit Dome.

Any island remediation project is only as good and as lasting as its environing physical conditions allow. Indeed, restorations are inherently impermanent, as island conditions are always changing and evolving. Even Runit Dome, everlasting though it was intended to be, is surrounded by ocean waters that, over the years of climate change, have risen—a phenomenon unforeseen by the dome's original designers, engineers, and builders.

In 1980, with the dome complete, the US government allowed the Native residents to return to their island homes. They came back willingly, and today, some 650 Marshallese live on two of the main islands of Eniwetok Atoll: Enjebi at the north, and Eniwetok Island at the south. Runit Island itself, still radioactive, remains uninhabited.

The question is whether the atoll, or any part of it, will be safe for much longer. In 1982, soon after the dome was completed, a National Research Council task force suggested that the structure could be breached by a severe typhoon. In the years since, various US agencies have conducted several additional studies of the dome's physical integrity and have assessed what the consequences might be of a major failure of the structure.

A recent evaluation of Runit Dome was released by the US Department of Energy in June 2020. Its *Report on the Status of the Runit Dome in the Marshall Islands* was a twenty-three-page document, written anonymously and submitted as a report to Congress. The text offered three main conclusions. First, "Runit Dome is not in any immediate danger of collapse or failure, and the exterior concrete covering the containment structure is still serving its intended purpose."

Second, the main risk posed by the dome was the flow of contaminated groundwater from the containment structure into the local marine environment. To evaluate the nature and the degree of risk involved, the Department of Energy (DOE) was in the process of establishing a groundwater radiochemical analysis program. The program would provide scientific data that would remove the uncertainties regarding "what, if any, effects the dome contents are having, or will have, on the surrounding environment now and in the future."

Third, radiological data already collected by the DOE's Marshall Islands Program "indicate that radiation dose rates to individuals on Eniwetok from internal exposure to fallout radionuclides are well below international standards for radiological protection of the public. . . . There is no evidence to suggest that the containment structure represents a significant source of radiation exposure relative to other sources of residual radioactive fallout contamination on the atoll."

But what all this really added up to was very much an open question. On the one hand, Runit Dome was still standing, relatively healthy, and doing the job for which it had been intended. Still, there was no getting around the fact that the dome housed an underground nuclear waste dump in an environment where the outflow of contaminated groundwater posed an increasing risk of further radiation exposure to the marine life in the lagoon and to the Native human population on the islands surrounding it. At this point, then, it is impossible to tell whether the radiological cleanup of Eniwetok Atoll was a success or a failure.

* * *

And then, finally, there was the artificial atoll and its largely synthetic focal point, Johnston Island. Surprisingly, in view of its having been given over to a succession of noxious uses throughout its history, the fact is that the restoration of this manufactured island was far and away the greatest success story of them all.

In 1970, shortly after the departure of the last POBSP fieldworker (Philip C. Shelton) from the atoll, Congress redefined Johnston Island's core mission. It would now become a site for the safe storage and disposal of chemical weaponry. Since 1917, the United States had produced an immense inventory of chemical warfare agents, and stored them in bulk containers as well as in individual munitions. The latter included shells, bombs, cluster bombs, bomblets, mortars, and various other projectiles, plus rocket warheads and land mines. Over the years, the Army Chemical Corps stockpiled these chemicals and devices at production plants and arsenals at several locations within the United States and, during and after World War II, at additional sites in Germany, Japan, Vietnam, and elsewhere.

In 1971, Johnston Island *also* became a chemical weapons storage facility. The island, being hundreds of miles from anywhere else, was the perfect place to house such munitions in total seclusion and relative safety.

And so an enormous quantity of chemicals—including the lethal nerve agents sarin and VX, as well as mustard gas—began arriving en masse upon Johnston Island. At its height a grand total of 31,496 tons of chemical agents were stored on the island in 412,732 individual munitions and bulk containers. In 1973, the army discovered that a number of barrels containing Agent Orange stored out in the open were leaking onto and into Johnston Island's sands and seeping out into the lagoon.

All of this while Johnston Atoll was still officially a national wildlife refuge.

By 1981, however, the Army's Chemical Corps had developed a plan for ridding the island of its entire store of chemicals. And this time, they did not forget about the birds.

The plan was to build a chemical weapons destruction facility, the Johnston Atoll Chemical Agent Disposal System (JACADS), a complex, gigantic structure that was a marvel of engineering, plumbing, furnaces, and smokestacks. Inside the facility, workers disassembled munitions and channeled their deadly contents into incinerators where high heat turned the chemicals into a flow of harmless molecules. At the end of the process, the result was a stream of environmentally benign hot gases released through tall smokestacks into the atmosphere. Still, men visiting the island during the incineration period had to shave off their beards to be able to use gas masks effectively if a leak were detected.

To monitor the effects of the JACADS operation on the atoll's ecosystem, the Department of Defense employed two scientists to study the its birds and marine life for a six-year period before the start of the incineration process and continuing on until the end, plus a three-year period afterward. The scientists were Phillip Lobel, a professor of biology at Boston University, and Elizabeth Anne Schreiber, an ornithologist at the Smithsonian Institution's National Museum of Natural History. Known to all as Betty, her husband, Ralph W. Schreiber, was a POBSP ornithologist who himself had spent 137 days at Johnston Atoll in 1966 and 1968.

At the end of the chemical incineration project, which took ten years, beginning in 1990 and ending in 2000, Betty Anne Schreiber wrote a detailed, 307-page account of the breeding biology and seabirds of Johnston Atoll. Her conclusion was that the birds and the chemical weapons disposal system had existed together quite harmoniously. Indeed, the atoll's bird populations of all species had actually *increased* across the duration of the weapons destruction program.

"The wildlife refuge is in excellent shape," Schreiber reported. "We monitored bird population, nest success, survival of adults from year to year, chicks up until breeding age, egg size and adult weight. We also did a comparison of birds within the vicinity of smoke stacks or birds that could be exposed to any possible agent leaks, with birds outside that area and found no difference in health or behavior."

Phillip Lobel was also enthusiastic about the ecological consequences of the weapons disposal program. "The JACADS project is without a question a complete success," he said. "The reason the Army had so much success was because they spared no expense throughout the whole project. The Army facilities were all top-of-the-line and the Army followed every possible safety precaution to ensure the continued growth of the National Wildlife Refuge."

In 2000, with the weapons destruction complete, the army gutted and dismantled JACADS building and removed it from the island. They demolished all the other buildings and structures as well, removed the hazardous waste, and turned the island into little more than a runway. That process took another three years to complete.

By May 2004, all remaining personnel left the island, leaving the place deserted. The island was then turned over to the Office of Property Disposal of the US General Services Administration, which put it up for sale to the general public.

Johnston Island now became, in effect and in actual fact, a *real estate listing*. The General Services Administration posted an ad for the place on its website, gsa.gov.

Johnston Island
Further Property Description:
Property Address:
717 nautical miles Southeast of Honolulu, Hawaii
Johnston Island, HI 94668
Type of the Property: LAND
Potential Usage: ecotourism, wildlife refuge
Sale Status:
Sale Method: Written Auction
Dimensions and Size of the Property:
Acres: 625.81

Comments:

Johnston Atoll consists of four small man-made islands enclosed in an egg-shaped reef approximately 21 miles in circumference. The wildlife refuge on the Atoll is a habitat for 32 species of coral, 300 species of fish, the endangered sea turtle and Hawaiian monk seal, and 30 species of migratory birds. Johnston, the main island, is 1000 yards long and 200 yards wide. The deed will contain use restrictions because the atoll was used by the Defense Department for storage of chemical munitions and as a missile test site in the 1950's and 60's. The island can be used as a residence or vacation getaway but it does not have utilities or a water supply. The airstrip and golf course are closed.

By any criterion, this was an amazing development. Johnston Island, a tiny land mass that had been devoted almost exclusively to hazardous uses—to nuclear testing work, missile launches, biological weapons trials, and the destruction of chemical weapons—was now transformed into an uninhabited tropical paradise, a potential "ecotourism" attraction.

You could buy the place and own it yourself, personally. All it lacks is water, electricity, and coconuts.

11
Aftermath and Aftereffects

At the end of the Pacific Project, many of its scientists managed to find ways to continue their lives as island-hopping vagabonds in search of birds. Such a life was not merely a job, but rather a calling, and once you were engaged in it, you did not go gently into another line of work.

That was true of Roger Clapp, for example, who after leaving the Pacific Ocean Biological Survey Program (POBSP) kept on doing pretty much what he had been doing all along, which is to say, visiting assorted barren and isolated Pacific islands, observing the local bird populations, and writing up accounts of what he saw. The problem was getting funding for these jaunts, but it turned out that there were a surprising number of organizations that, for reasons of their own, took a special interest in birds, or were engaged in projects in which ornithology could be practiced as a sideline. And so when the US Fish and Wildlife Service did its own survey of the Northwest Hawaiian Islands, Roger Clapp went along to assist. And when the US Coast Guard left to perform its annual inspection tour of the lighthouses and day beacons in the Phoenix Islands and American Samoa, he again hitched a ride.

After that, Clapp recalled, "field work, to my disappointment, became much more limited but I managed to revisit some of the southern islands in 1970 and 1971 as a consultant to the U.S. Air Force, which wanted to put some towers on these islands to track ballistic missile launches into Canton Atoll lagoon."

Later, on a visit to Laysan Island in 1978, Clapp had the unusual experience of seeing a bird appear to commit suicide. "One young Black-footed Albatross was particularly unfortunate," he wrote in a 2001 memoir. "It apparently had a defect in its barbules making its feathers completely unable to support flight. Day-by-day it stood there, slowly starving to death, but I guess it eventually decided on suicide. I watched as it walked to the edge of the water, swam a few hundred yards out to sea, and was engulfed by a huge maw coming up from below, the tips of its wings being the last seen as the bird was dragged below."

These junkets didn't last forever, but Clapp pursued such opportunities for as long as he could. His last trip to the Pacific came in 1988, when he traveled to Kwajalein Atoll in the Marshall Islands, courtesy of the US Army Corps of Engineers, which wanted to survey the atoll's wildlife resources for an environmental impact statement pertaining to the army garrison stationed there. Clapp even managed to get a publication out of the trip: "Notes on the birds of Kwajalein Atoll, Marshall Islands," in the *Atoll Research Bulletin* (1990).

When not on his irregularly scheduled forays to the world's most remote and solitary islands, Roger Clapp held a virtually lifelong position at the Smithsonian's National Museum of Natural History, where he curated one of the largest bird skin collections in the world, with approximately 625,000 specimens.

"My desk stands less than 10 feet from where it did nearly 40 years ago," he said of his workplace, "although it now has two computers rather than a somewhat beat-up typewriter."

Still, in Roger Clapp's view, no amount of office work could bear comparison with being out on the water or walking the islands. "I still remember my years in the Pacific as the high point of my career, and only wish I could visit them once again."

The one thing he hated with a passion was Washington, DC, traffic. Being stuck in a long line of cars at a standstill was as far away as you could possibly get from the clear skies and open seas of the Pacific. Clapp lived in rural Virginia, in a town called Aldie, about forty miles from the District of Columbia. And so to avoid the rush hour commutes, he'd get up at five in the morning and start driving in, and in the afternoon he'd leave for home by 3:30. But even that routine had to end. Roger Clapp died on Christmas eve, December 24, 2018, at the age of eighty.

* * *

Binion Amerson always regarded the POBSP as "the last big exploratory adventure," and he himself was the chronicler of much of it. During his tenure as research curator for the POBSP, Amerson was the author or coauthor of at least sixteen monographs or papers on the islands, atolls, and bird and insect species of the Pacific. Two of these reports—one on the French Frigate Shoals and the other on the Marshall and Gilbert Islands—were of such length as to qualify as books in their own right. Indeed, both were published as an entire, standalone volumes of the journal in which they appeared, the *Atoll Research Bulletin*.

Amerson had a particularly strong and somewhat puzzling affinity for the French Frigate Shoals. Not only did he write the *ARB* account of the place, in 1967 he even wrote a movie script entitled *French Frigate Shoals: The True Center of Hawaii*. And there is today a rather poor quality, fifty-seven-minute-long YouTube video, produced by Amerson, about them. In the late 1990s, finally, Amerson wrote *The Coral Carrier*, a book about Tern Island in the French Frigate Shoals that navy Seabees had turned into the airport that resembled an aircraft carrier. He went so far as to have this work privately printed and published. "The Coral Carrier sails on," was its refrain.

In the immediate aftermath of his seven years with the POBSP, Amerson returned to the University of Kansas to finish work on his master's degree in ecology and environmental studies. His thesis, completed in 1973, drew upon his POBSP work in the Northwestern Hawaiian Islands.

Thereafter, he put his writing talents to use at Environmental Consultants, of Dallas, for which he produced a two-volume work on the ecology of American Samoa, a study commissioned by the US Fish and Wildlife Service. During the 1980s, the demand for such ecological reports dropped to the point that Amerson moved into work as a technical writer in the information technology industry.

He lived alone in a small house in the Farmers Branch suburb of Dallas where, one morning in the spring of 1994, he visited a shopping mall in the area. A daylily show was in progress there, and Binion was struck by the color and variety of the flowers. He soon became an avid breeder of daylilies himself. Before moving out of his home in 2011, he donated more than seven hundred daylily cultivars to the Farmers Branch Public Daylily Garden, where his collection remains today. Amerson died on September 23, 2017, and donated his body to science.

* * *

During his two years with the Pacific Project, Fred Sibley helped band a grand total of 150,000 birds, which was an impressive number by any measure. "But that's the whole crew, three to five people on eight different two-month cruises," Sibley says today. "On one trip we did 60,000. This high number was totally dependent on the presence of a Sooty Tern colony. With millions of birds in the colony we did 10,000 one night."

In 1965, amid rumors that the project was going to end because of loss of funding (which didn't happen), and with an opportunity to join the condor research center of the US Fish and Wildlife Service in California, Sibley left the POBSP and accepted the job. He was an endangered species biologist there, with a special interest in saving the nearly extinct California condor.

Sibley spent four years with the Fish and Wildlife Service but left in 1969 to become director of the Point Reyes Bird Observatory, where he believed he could do more for bird conservation in general. A year later, in 1970, the tectonic plates shifted once again when his old teacher at Yale, Charles Sibley, became director of the Peabody Museum of Natural History at Yale, the very place where Dillon Ripley had earlier, from 1959 to 1964, been director.

"Charles Sibley hired me as collection manager when I had four kids of school age and desperately needed a permanent job, and ideally one involving field work," Fred Sibley recalled. He started as collection manager for ornithology and later became manager for all vertebrate collections. Fred Sibley was quite happy working with Charles, and the two Sibleys published a number of papers together.

After almost twenty years at the Peabody Museum, Fred Sibley retired in 1998. He moved to Alpine, New York, where he lives today, in the house he grew up in as a kid.

* * *

Larry Huber had the shortest post-POBSP career of them all.

He had been in the program as a research assistant from 1963 to 1969. During that time, he wrote a number of internal POBSP reports. After leaving the project, he returned to Tucson and he married a woman named Marguerite.

He contributed one paper, "Notes on the Migration of the Wilson's Storm Petrel," a small, migratory seabird, to a journal called *Notornis*, which is a quarterly publication of the Ornithological Society of New Zealand. It was published in March 1971. At the time he wrote the piece, Huber was in the process of writing a joint paper with John Bushman, of the Deseret Test Center, about the avifauna of Eniwetok Atoll. But he did not get to finish it, and the article never appeared.

Later in 1971, Huber and his wife, together with their friends Louis and Josephine Petit, were in a small, open boat collecting bird eggs on some islands off the Baja California coast. The facts beyond that are sketchy. Richard Crossin, who was friendly with Huber during the program, says that their boat was swamped in a storm, and all four drowned. The date was July 11, 1971.

"A search plane flew over the area and saw the overturned boat and two bodies," Crossin recalled. "In the end, only one body was recovered." Jane Church, "den mother" to the POBSP fieldworkers (who herself died November 2016, in Winchester, Virginia) thought it was Marguerite, but this is not certain. There is a black and white picture in the Smithsonian POBSP archives relating to the incident. It shows a black body bag, but there is no indication as to whose body it contained.

And thus ended the short and otherwise happy life of Lawrence N. Huber, who was twenty-seven at the time of his death.

* * *

Although they were not officially a part of the Pacific Program, Project SHAD and its participants were in some ways inseparable from it. The POBSP and SHAD had a common origin in Project 112, the weapons assessment program that gave rise to them both. Further, both POBSP and SHAD were directed, in whole or in part, by the Deseret Test Center. The personnel of the two programs often traveled together, aboard the same ships.

Both the POBSP fieldworkers and SHAD personnel faced a range of dangers. The POBSP fieldworkers made risky landings on islands and atolls in high surf, they suffered bites and infections during their bird banding work, plus bouts of seasickness, sunburn, and brushes with tropical diseases. On his first visit to the Phoenix and Line Islands, Roger Clapp was bitten so many times by masked boobies while banding them that "by the time the two months of field work were completed, I had five major infections on one hand, four on the other, and my hands were swollen so badly that that they flew me back from Canton, rather than let them fester during the six or seven days it would take to return to Honolulu [by ship]."

But for all the hazards they faced and surmounted, the fieldworkers emerged from the program in relatively good physical condition, and many went on to live long and healthy lives.

Not so some of the SHAD crew members. Most of these men confronted even greater apparent threats in the form of the lethal chemicals and pathogenic biological agents that were repeatedly sprayed over their ships. Although few of the working seamen knew it at the time, the boats on which they traveled were doused with chemical nerve agents such as VX and sarin, substances that are lethal to humans even in small quantities. Their ships also steamed through clouds of pathogens such as the Q fever microbe (*Coxiella burnetii*), the tularemia microbe (*Pasteurella tularensis*), and the bacterial toxin staphylococcal enterotoxin B. It would not be surprising, therefore, if a certain subset of the full SHAD population sooner or later developed health problems that might be linked to possible exposures to these and other disease-causing agents.

And as the years passed, a number of Project SHAD veterans did indeed suffer illnesses that they themselves attributed to their unique shipboard experiences. And who could blame them? The men who took part in the Shady Grove tests knew that ranks of caged monkeys were arranged on the open decks of their ships and exposed to experimental agents while the sailors themselves were in closed-off "safety citadels." They also knew that afterward the monkeys were transferred to the *Granville S. Hall* laboratory ship while the light tugs that had passed through the cloud were being washed down and otherwise decontaminated by men wearing gas masks and other protective gear.

The secrecy surrounding the whole operation also fostered doubts about their own roles in the project. Were they themselves unwitting guinea pigs? If so, it would be logical to think that their health would be negatively affected at some point in the indefinite future.

There was the case of Jim Sohns, for example, a retired Honolulu police officer who was a ship's storekeeper on the *Granville S. Hall* during the Shady Grove series of bioweapons trials at sea. He was also on the ship during two other tests, Flower Drum, in which the nerve agent sarin was used, and Fearless Johnny, a test of the lethal VX chemical. In 2002, after reading some recently released Department of Defense fact sheets on the tests, Sohns was sure that one or another of the substances was the cause of his medical problems.

"They [the tests] ruined my life," he told a reporter for the *Honolulu Star Bulletin* after the DOD fact sheets came out. "It destroyed my marriage. I lost my sex drive five years ago. I couldn't explain why to my wife. I no longer can do anything. I can't walk more than 50 to 100 feet without running out of breath. I had to buy a scooter to get around."

And Sohns was not alone in his complaints. Jack Alderson, the tugboat fleet commander, suffered diminished lung capacity and had prostate cancer. Robert W. Bates was in Operation Eager Belle II, a test of *Bacillus globigii*, the anthrax simulant, and contracted pneumonia a little while afterward. George J. Brocklebank was in a SHAD test called Copper Head, in which *Bacillus globigii* and the tracer chemical zinc cadmium sulfide were dispersed, and he later developed circulatory problems in his legs.

The problem with these and other such cases, however, was that there was no sure way of connecting the illnesses of a given SHAD veteran to the weapons trials they were in or to the substances involved. People develop health problems all the time, more so as they aged; that was common knowledge and a fact of life.

Still, the Department of Defense Fact Sheet about the Flower Drum test series that Jim Sohns participated in made the tests sound hazardous enough. "The USS *George Eastman* (YAG-39) was exposed to candidate sarin nerve agent simulant and to the actual agent. The ship was enveloped by the test agent disseminated by a gas turbine mounted on the bow of the ship and by simulated envelopment—direct injection of the test agent [simulant] into the air supply system."

But there was ample reason to think that the illnesses suffered by the SHAD crew members were not a product of contact with the agents used in the tests. If sailors aboard the *George Eastman* had been directly exposed to either sarin or VX during the Flower Drum or Fearless Johnny tests, the ship would have returned full of dead men; the two agents were that lethal.

There were two reasons why the *George Eastman* crew members were unlikely to have been directly exposed to sarin. The first is that during the tests, all crew members wore protective masks, while members of the dissemination crew also wore protective clothing. Second, personnel who were not part of the dissemination crew were confined to the safety citadel for the duration of the test.

The Flower Drum test series occurred in two different phases, both off the coast of Hawaii. Jim Sohns was in the Phase I test series, which took place in the months February through April, and again in August through September 1964. A partially declassified 1965 report on the Flower Drum Phase I test series describes the safety precautions in place during the trials: During dissemination of sarin nerve agent, "the disseminator crew wore M5 protective ensembles and all other personnel (who were in the safety citadel) wore MK5, M7A1, or M17 protective masks. After dissemination, all personnel whose duties required them to leave the safety citadel wore protective masks until the ship was cleared of [sarin]. During the dissemination period of the simulant trials, all personnel wore protective masks."

In the particular case of Jim Sohns, there was yet an additional and special reason why he was unlikely to have been affected by any of the agents used during the Flower Drum, Phase I tests: he was not on the ship that had been exposed to sarin, the *George Eastman*, but was rather aboard the laboratory ship, the *Granville S. Hall*, which participated as an escort and laboratory facility. The final report on the Phase I test series states that the USS *George Eastman* "steamed into the wind maintaining a relative wind speed of 10 to 30 knots. The [USS *Granville S. Hall*] maintained a parallel course forward and starboard of the [USS *George Eastman*]."

The *Granville S. Hall*, in other words, sailed *in front of* the *George Eastman*, which means that only the *Eastman*, and not the *Hall*, was exposed to either the sarin agent or the simulant.

Nevertheless, there were plenty of other SHAD personnel besides Jim Sohns who participated in several biological and chemical weapons tests in the Pacific and came forward with medical complaints in the years afterward. Anecdotal reports of illnesses on the part of a subset of participants do not by themselves constitute evidence of causality, however. Correlation is not causation. And so the only way of establishing a cause-and-effect relationship between the SHAD tests and negative health outcomes on the part of participant crew members was by performing an epidemiological investigation of the incidence of disease across the entire SHAD cohort.

Which was in fact done. Indeed, it was done twice.

* * *

In 2002, the Department of Veterans Affairs asked the Institute of Medicine to undertake an epidemiological study of the potential long-term health effects of participation in Project SHAD.

The Institute of Medicine (IOM) is part of the National Academies of Sciences, Engineering, and Medicine, a private nongovernmental institution that was established in 1863, during the tenure of President Abraham Lincoln, to advise the nation on issues relating to science and technology. As such, it was one of the country's most highly respected institutions of scientific research, and was eminently qualified to do the work proposed. The study was no small undertaking, and would be conducted over a three-year period at a cost of $3 million.

The IOM assembled an elite group of experts in medicine, psychiatry, and public health, and appointed a project staff of thirteen researchers. Using information provided by the Department of Defense, the researchers compiled as complete a list as they could of the army, navy, and marine corps personnel who participated in Project SHAD. Then, as a control group, the staff members also put together a list of nonparticipants against whom the health outcomes of the SHAD personnel could be compared.

The researchers ended up with a list of 5,867 Project SHAD participants and 6,757 nonparticipant controls. In the process of analyzing the data, the IOM staff members discovered that participation in the SHAD operation was not a single, unitary experience that was the same for all the servicemen.

Instead, the SHAD veterans could be divided into three main exposure groups. The first, by far the largest, consisted of more than three thousand men whose potential exposure was limited to one or another of two simulant agents: BG (*Bacillus globigii*), the anthrax simulant, and MAA (methylacetoacetate), a sarin nerve agent simulant.

The second group consisted of about 850 participants whose only potential exposure was to yet another simulant, TEHP, a chemical considered to be of little toxic risk. These two groups, a total of almost four thousand participants, had been potentially exposed to substances that were not considered to be hazardous at the time the tests were done, although some were later found to cause illness among immunocompromised subjects.

The third group was the smallest group of all. It consisted of about 720 participants in tests during which active biological or chemical agents were used. This was the personnel subset in which the risk of adverse health effects appeared to be highest and who therefore ought to be the sickest.

But when the IOM scientists compared the health outcomes of that group against its respective control group, they found that there was not much difference in their health statuses. In its final report, a 143-page document, *Long-term health effects of participation in Project SHAD*, released in 2007, the Institute of Medicine's scientific panel concluded that "in general, there was no difference in all-cause mortality between Project SHAD participants and nonparticipant controls, although participants statistically had a higher risk of death due to heart disease. . . . Participants also reported statistically significantly worse health than controls, but no consistent, specific, clinically significant patterns of ill health were found." Overall, "we have found no clear evidence of specific health effects that are associated with Project SHAD participation."

There was a caveat, however: "We must remark that this does not constitute clear evidence of a lack of health effects." Proving a *lack* of adverse health effects would be difficult, however, as it would amount to proving a negative, which is notoriously problematic. Although it could be done.

* * *

It is safe to say that the Institute of Medicine's report was not a big hit with the SHAD veterans themselves. In June 2008, Jack Alderson, who had emerged as a spokesman for the group, appeared on the *CBS Evening News* with Katie Couric. He went down a list of his fellow officers. "McFadden is dead. McQueen I can't find. Smith I can't find. Forster's dead. Goforth's dead. And they died of respiratory problems, lung problems. I've had a malignant melanoma—a big one. I have prostate cancer."

Various congressional members who volunteered to speak for for the SHAD crews called for a second epidemiological study. And in 2010, Congress authorized it. The new investigation would also be conducted by the Institute of Medicine, and it would prove to be an even more sweeping and ambitious undertaking, pursued in greater detail, with greater thoroughness, and with a fuller access to declassified documents than the earlier study had had. And at the end of it, in 2016, the scientific team issued an even longer, 196-page report, *SHAD II*. Its conclusions, however, were even more clear cut that the health of the SHAD servicemen was not compromised by their participation in the Project SHAD trials at sea.

In short, the investigative team discovered that, "in finding no overall differences in all-cause mortality between SHAD participants and comparison groups, the SHAD II analysis agreed with the results of the previous IOM study. However, it differed in that, with an additional 7 years of follow up, it did not find an elevation in heart disease mortality. Indications of an increased risk of heart disease in the crew of the USS *George Eastman*, one of the exposure groups examined, did not attain statistical significance after adjustment for the multiple comparisons carried out in the analysis. The SHAD II analysis found no clear differences in degree of illness between SHAD participants and the comparison group."

The scientists acknowledged that they could not rule out a possible negative health effect with absolute certainty. Such certainty did not lie within the province of science. "However," they said, "within the limits of the data available to the committee, the results of the analyses provide no evidence that the health of SHAD veterans overall or those in the exposure groups is significantly different from that of similar veterans who did not participate in these tests."

There was also the intriguing further fact that in many categories of illness, Project SHAD veterans had a slightly *lower* (albeit statistically insignificant) incidence of disease than corresponding members of the control group, almost as if being a SHAD participant might have had a slightly protective health effect. On a strictly scientific basis, therefore, there was no longer any documented evidence that participation in Project SHAD resulted in any negative health effects.

* * *

And then, finally, there were the effects of the POBSP on the Smithsonian Institution itself.

The Smithsonian's connection to the American germ warfare program was revealed to a mass audience by NBC's *First Tuesday* program in February 1969. Although it seemed like major news at the time, in the end it turned out to have little staying power. The year 1969, after all, was the year of the moon landing (July), as well as the first flight of the Boeing 747 (February), the first flight of the Concorde (March), and the iconic Woodstock rock concert (August). In the face of that train of events, news about the Smithsonian–army connection was merely a nine-day wonder, soon enough forgotten.

But the whole scandal was resurrected again, sixteen years later, in May 1985, when the *Washington Post* ran a new account of the affair as the cover story of its Sunday magazine section. Entitled "The Smithsonian's Secret Contract: The link between birds and biological warfare," it was written by longtime investigative journalist Ted Gup.

"The Pacific project was two separate missions existing side by side, the Smithsonian's and the Pentagon's," Gup wrote. "The Smithsonian was only too eager to be given funds to study bird migratory patterns and the military was eager to find 'safe' sites for atmospheric testing of biological weapons in the Pacific."

That was correct, and it described the project in a nutshell. It was also correct in saying that the program "was one of the largest and most mysterious undertakings in the institution's 139-year history."

But the piece also hinted darkly of "Strangelovian fantasies when even one of God's gentlest creatures, a gull, could be considered for a doomsday assignment." This was the "bird bomb" hypothesis surfacing yet again. But Gup provided no evidence that anyone in the military, or at the Smithsonian, had ever taken the idea seriously. Gup noted that in 1961 the CIA had funded a project called "Role of Avian Vectors in the Transmission of Disease," but that title could apply equally to the natural transmission of disease and deliberately engineered, malicious transmission. And since the project was classified, there was no way for the public to know which meaning was intended. But whichever the title meant, the fact that the CIA might have been considering the use of birds as intentional vehicles of disease does not mean that the army was, much less anyone at the Smithsonian.

In any case, none of the *Washington Post*'s belated rediscovery of "the Smithsonian's secret" seemed to have made much of an impact one way or the other. The institution's own Senate of Scientists had long ago, in 1969, closed its books on the Pacific Project and never looked back. Later, when a new generation of Smithsonian scientists wrote or spoke about the Pacific Project, which was infrequently enough, they portrayed it in the warmest of terms. For example, in 2015, at a World Seabird Conference held in Cape Town, South Africa, Smithsonian ornithologist Autumn-Lynn Harrison gave a talk about the project entitled, "An ode to the Pacific Ocean Biological Survey Program and its biologists, 1963–1969." In general, an ode is a hymn of praise.

Still, it is arguable that while the Pacific Project was at its height, the Smithsonian and Secretary S. Dillon Ripley were just then in the process of making an even worse mistake than participating in the POBSP was in the eyes of its critics. And this was the founding of what would turn out to be the institution's least-visited, least-distinguished, and deservedly the most un-known branch, the Anacostia Neighborhood Museum.

The idea for it went back to 1966, when Dillon Ripley hatched a plan to put "an experimental store-front museum" in a Washington, DC, neighborhood that was about three miles away from the National Mall. Ripley's goal was to reach out to an African American community to encourage residents to visit the institution's other, primary museums. There was nothing inherently bad about the idea; the flaw was in its execution.

The neighborhood chosen was Anacostia, a predominantly Black commu-nity located in Southeast Washington, near the Anacostia River, after which the neighborhood was named. The site selected was the Carter Theater, a former movie house that had opened for business in 1948, only to close its doors twelve years later, in 1960.

The Smithsonian acquired the building in March 1967 and renovated the interior but kept the outdoor marquee. This meant that when it opened to the public in September 1967 as the Anacostia Neighborhood Museum, it still looked like a movie theater. Under the marquee, a small sign bore the legend "Smithsonian Institution" not carved in stone, as was more usual for a proper Smithsonian edifice, but rather just painted, in black script lettering, on a small, barely visible signboard (fig. 11.1).

Fig. 11.1 Anacostia Neighborhood Museum
Credit: Anacostia Community Museum, Smithsonian Institution

An early exhibit, which ran from November 1969 to January 1970, was about rodent infestations in urban neighborhoods: *The Rat: Man's Invited Affliction*. Misbegotten from the start, the Anacostia Neighborhood Museum had nowhere to go but up, which it did, although it took a while. By 1987 a new and impressive structure was designed and built, during the tenure of Ripley's successor, Robert McCormick Adams. It opened in April of that year as the Anacostia Museum (later renamed the Anacostia Community Museum), ten blocks from the original, and was a success.

In 1971, meanwhile, C. Dillon Ripley, fresh from his movie theater/museum fiasco, hired the former Apollo 11 astronaut Michael Collins to oversee the design and construction of another new museum building on the Mall in Washington, the National Air and Space Museum.

"I will be very disappointed, and surprised, if it does not turn out to be the most exciting museum in the world," Collins said during the construction phase.

He needn't have worried. The museum opened on schedule and under budget, on July 1, 1976, in the nation's bicentennial year, and it soon became one of the most visited museums of any type, anywhere in the world. Dillon

Ripley thereby became the holder of the unique distinction of having been responsible for the creation of both the least- and most-visited museums of the Smithsonian Institution. In 2018, the Air and Space Museum had more than six million visitors.

By that time, the Smithsonian Institution as a whole had become the world's largest museum and research complex. It included nineteen museums, twenty-one libraries, nine research centers, and a zoo. Collectively, they hosted thirty million visitors annually. Such an institution more than fulfilled the long-ago hopes of its original benefactor, James Smithson.

If the Smithsonian gained some of its fame, holdings, and knowledge in unorthodox ways, that was a minor failing. The institution would survive other controversies, such as the 1994 exhibit of the *Enola Gay*, the B-29 Superfortress that dropped the first atomic bomb on Hiroshima. The exhibit was accompanied by interpretive commentaries that some critics charged unfairly politicized the event and cast a bad light upon American airmen. The Smithsonian responded by removing the commentaries in question and replacing them with strictly technical and historical data. That move, too, had its critics, and the exhibit was cancelled in January 1995. The National Air and Space Museum opened a modified version of the exhibit in June 1995, and it ran until May 1998.

The year 2021 was the 175th anniversary of the founding of the Smithsonian Institution in 1846. To commemorate the occasion, the institution installed a new exhibition, *FUTURES*, in the Arts and Industries Building, which was opened in 1881. A Smithsonian press release said, "'FUTURES' will spot-light the Institution's historic role as an engine of the future. Since 1846, the Smithsonian's mission for the 'increase and diffusion of knowledge' has led to remarkable leaps—exploring the beginnings of the universe, saving species from extinction, preserving the full diversity of human culture and bringing new methods of digital learning to billions around the globe."

The exhibit did not exclude artifacts from the past, so long as they looked toward the future: on display, for example, was a Planetary Society space sail for deep space travel, as well as the world's first controlled thermonuclear fusion device (which created heat but did not explode). The exhibit did not in any way acknowledge the POBSP, which was very much in the past. A few relics of the Pacific Project were on public view, mainly bird skins mounted in the ornithological display cases of the National Museum of National History.

But the program itself was best forgotten and passed over in silence.

Epilogue

On the morning of January 25, 1904, James Smithson—or at least what little was left of him—finally arrived in America. As was true of many of the events that pertained to Smithson, both during his lifetime as well as for several years after his death, the transfer of his bones to US shores was not without its proper share of drama.

Over the previous seventy-five years, the body of James Smithson, who had died in 1829, had lain in a grave in the San Benigno Cemetery, located high on a bluff overlooking the harbor at Genoa. The cemetery itself was owned by the British, but the ground underneath it, all the way down to sea level, was owned by an Italian marble quarry. The quarry operators were slowly blasting the cliff away, and by 1900 they were approaching the cemetery's border. In that same year, the quarry's owner informed the British consulate that by 1905 their excavation would reach, undermine, and ultimately destroy the cemetery, sending down the hillside whatever buried coffins remained there.

Word of Smithson's involuntary exhumation soon reached the Smithsonian Institution, whereupon its governing Board of Regents took up the question of bringing the founder's mortal remains to Washington. And, just as the US government had been initially diffident about accepting Smithson's money, so too were the institution's regents about accepting Smithson's bones. In 1901, when the question was put to a vote, only a single member of the Smithsonian's board was in favor. And that member was Alexander Graham Bell.

At that point Bell took the matter into his own hands. He decided to travel to Europe with his wife, Mabel, who was deaf, take possession of Smithson's remains, and bring them back to the States. After all, Bell, who was fifty-six, had been advised by his physician to rest and to take an ocean voyage for the sake of his health. Besides, he naively imagined that bringing back the bones would be a relatively easy matter. It wasn't.

In 1903 Alexander Graham Bell and his wife sailed for Paris, from which city they would take a train to Genoa. He caught a cold on the train from Paris, and by the time the couple arrived in Genoa, on Christmas day, it was cold and raining, and Bell was miserable.

Still, that was hardly the worst of it. Several barriers to the exhumation and transfer process lay ahead. "There seemed to be no end to the red tape necessary to remove the body," Mabel wrote in her journal. "A permit to export the body beyond Italian limits, a permit to open the grave, a permit to purchase a coffin, permits from the National government, city government, the police, etc., etc."

And at length, just as all those obstacles were overcome, a new one presented itself with the discovery that the de la Batut family had long ago issued a permanent injunction against the removal of the body. However, by offering bribes and other inducements to the appropriate Italian bureaucrats, Bell prevailed, had the grave opened, the bones removed, and put into a new, zinc coffin.

On January 7, 1904, the coffin containing Smithson's bones, which were in moderately good shape, having separated from each other but not having crumbled into dust, were loaded onto the steamship *Princess Irene*. The Bells also boarded the ship, which then sailed for New York.

They arrived in New York Harbor on January 24, 1904, but did not set foot on land. Instead, Bell and the coffin were transferred directly to the US Navy dispatch vessel USS *Dolphin*, which soon left for Washington.

James Smithson's mortal remains first arrived on US soil when the boat steamed into Washington Navy Yard the next day. Early on the morning of January 25, 1904, Smithson's casket was lifted from the *Dolphin*, deposited on the dock, and carried to the Navy Yard gate. James Smithson, at long last, had finally come to America.

By February 1904, the skull and bones of James Smithson were installed in the Smithsonian Castle, where, in a crypt, they lie today.

* * *

The Johnston Island real estate listing, the one that in 2005 offered the island to the general public as "a residence or vacation getaway" and as a possible "ecotourism" destination, didn't last very long. But not because the property sold.

The island had been put up for sale by Rock A. Southward, who was a realty officer at the General Services Administration's Office of Property Disposal, in San Francisco. Southward posted the webpage listing for the island as a "teaser," which was evidently a common practice for the GSA, employed to measure possible public interest in a somewhat unusual piece of government real estate. Southward said later that he deleted the listing after a few days.

Apparently, he had received a few tentative bids for the place, but he did not say what the actual dollar amounts were or why none was accepted.

After being withdrawn from the real estate market, Johnston Island remained under US Air Force ownership and control. But the air force wanted nothing more than to be rid of the place, which by this time had become a white elephant, a site that nobody really wanted or had any good use for. As it happened, that newfound uselessness would turn out to be its salvation.

In 2009, the status of Johnston Atoll was finally resolved when President George W. Bush formally proclaimed it, together with Wake, Howland, Baker, and Jarvis Islands, Palmyra Atoll, and Kingman Reef, the Pacific Remote Islands Marine National Monument. The purpose of the monument was to preserve the marine environment and terrestrial life forms of the US outlying minor islands and atolls. The proclamation stated, in particular, that the relevant government authorities "shall not allow or permit any appropriation, injury, destruction, or removal of any feature of this monument except as provided for by this proclamation and shall prohibit commercial fishing within boundaries of the monument."

In other words, from that time forward these islands really would be treated as actual nature preserves, as special, as places where birds, mammals, and other life forms would be left in peace and could thrive and reproduce without fear of human intervention. There would be no further nuclear testing, or biological weapons trials, or chemical weapons storage or disposal, or any other such historically permitted—and indeed common— practices, within the areas covered by the refuge. Instead, the protected islands and atolls would at long last be left alone, given back to themselves, allowing them to return to conditions resembling their prediscovery, natural, and native states.

For the wildlife of those areas, this was progress.

APPENDIX I

The original contract creating the POBSP

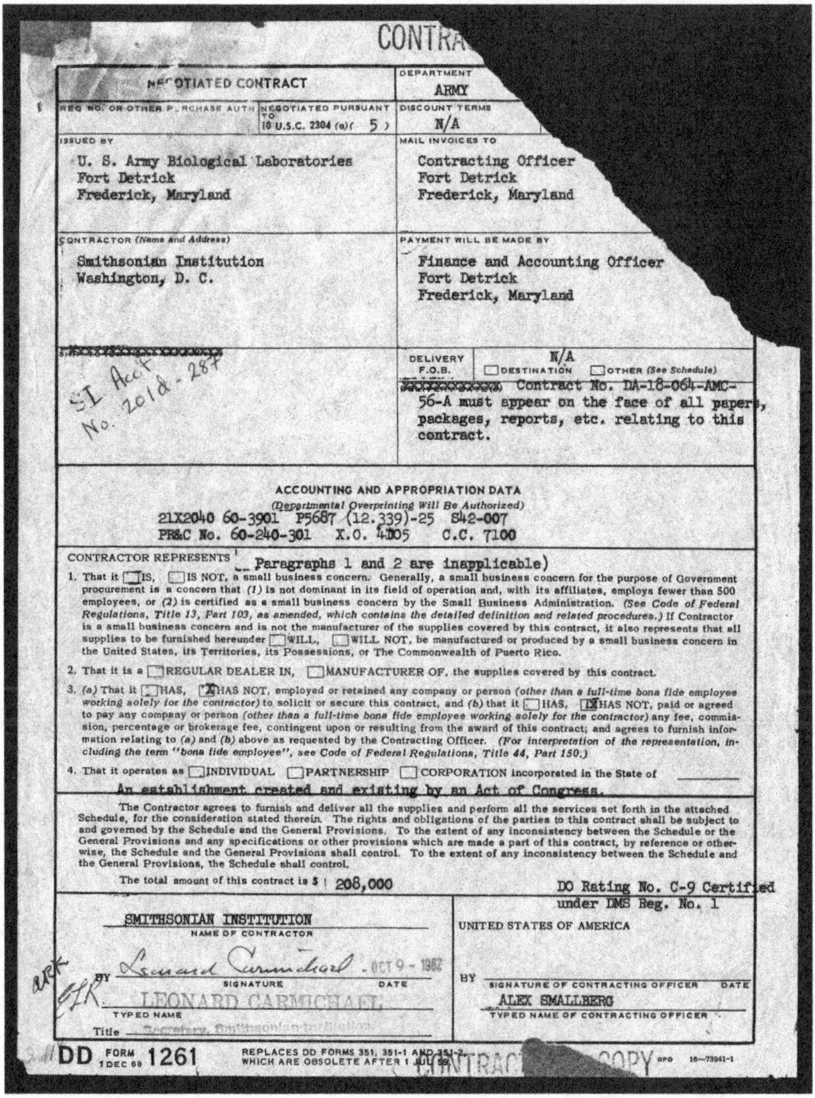

CONTRA[…]

NEGOTIATED CONTRACT

DEPARTMENT: ARMY

REQ. NO. OR OTHER PURCHASE AUTH.

NEGOTIATED PURSUANT TO: 10 U.S.C. 2304 (a)(5)

DISCOUNT TERMS: N/A

ISSUED BY:
U. S. Army Biological Laboratories
Fort Detrick
Frederick, Maryland

MAIL INVOICES TO:
Contracting Officer
Fort Detrick
Frederick, Maryland

CONTRACTOR (Name and Address):
Smithsonian Institution
Washington, D. C.

PAYMENT WILL BE MADE BY:
Finance and Accounting Officer
Fort Detrick
Frederick, Maryland

XXXXXXXXXXXXXXXX

SI Acct No. 2012-287

DELIVERY F.O.B. N/A ☐ DESTINATION ☐ OTHER (See Schedule)

XXXXXXXXXX Contract No. DA-18-064-AMC-56-A must appear on the face of all papers, packages, reports, etc. relating to this contract.

ACCOUNTING AND APPROPRIATION DATA
(Departmental Overprinting Will Be Authorized)
21X2040 60-3901 P5687 (12.339)-25 S42-007
PR&C No. 60-240-301 X.O. 4D05 C.C. 7100

CONTRACTOR REPRESENTS (Paragraphs 1 and 2 are inapplicable)

1. That it ☐ IS, ☐ IS NOT, a small business concern. Generally, a small business concern for the purpose of Government procurement is a concern that *(1)* is not dominant in its field of operation and, with its affiliates, employs fewer than 500 employees, or *(2)* is certified as a small business concern by the Small Business Administration. *(See Code of Federal Regulations, Title 13, Part 103, as amended, which contains the detailed definition and related procedures.)* If Contractor is a small business concern and is not the manufacturer of the supplies covered by this contract, it also represents that all supplies to be furnished hereunder ☐ WILL, ☐ WILL NOT, be manufactured or produced by a small business concern in the United States, its Territories, its Possessions, or The Commonwealth of Puerto Rico.

2. That it is a ☐ REGULAR DEALER IN, ☐ MANUFACTURER OF, the supplies covered by this contract.

3. *(a)* That it ☐ HAS, ☒ HAS NOT, employed or retained any company or person *(other than a full-time bona fide employee working solely for the contractor)* to solicit or secure this contract, and *(b)* that it ☐ HAS, ☒ HAS NOT, paid or agreed to pay any company or person *(other than a full-time bona fide employee working solely for the contractor)* any fee, commission, percentage or brokerage fee, contingent upon or resulting from the award of this contract; and agrees to furnish information relating to *(a)* and *(b)* above as requested by the Contracting Officer. *(For interpretation of the representation, including the term "bona fide employee", see Code of Federal Regulations, Title 44, Part 150.)*

4. That it operates as ☐ INDIVIDUAL ☐ PARTNERSHIP ☐ CORPORATION incorporated in the State of ______
An establishment created and existing by an Act of Congress.

The Contractor agrees to furnish and deliver all the supplies and perform all the services set forth in the attached Schedule, for the consideration stated therein. The rights and obligations of the parties to this contract shall be subject to and governed by the Schedule and the General Provisions. To the extent of any inconsistency between the Schedule or the General Provisions and any specifications or other provisions which are made a part of this contract, by reference or otherwise, the Schedule and the General Provisions shall control. To the extent of any inconsistency between the Schedule and the General Provisions, the Schedule shall control.

The total amount of this contract is $ 208,000

DO Rating No. C-9 Certified under DMS Reg. No. 1

SMITHSONIAN INSTITUTION
NAME OF CONTRACTOR

BY Leonard Carmichael OCT 9 - 1962
SIGNATURE DATE

LEONARD CARMICHAEL
TYPED NAME

Title Secretary, Smithsonian Institution

UNITED STATES OF AMERICA

BY
SIGNATURE OF CONTRACTING OFFICER DATE

ALEX SMALLBERG
TYPED NAME OF CONTRACTING OFFICER

DD FORM 1 DEC 58 1261 REPLACES DD FORMS 351, 351-1 AND 351-2, WHICH ARE OBSOLETE AFTER 1 JUL 59. CONTRACT COPY

Fig. A.1 [TK, December 2021] The original contract creating the POBSP

Credit: Smithsonian Institution Archives, RU 245 Box 8, Folder 20

Entomological Warfare

The US biological warfare program can be said to have begun on November 18, 1941, when a group of nine of the nation's top biologists, from places such as Johns Hopkins, Yale, and the Rockefeller Institute, gathered for a meeting at the National Academy of Sciences in Washington. They worked under the title of the WBC committee, which officially meant "War Bureau of Consultants," but unofficially was a deliberate backwards spelling of "Committee on Biological Warfare." The purpose of the meeting was to assess whether bacterial warfare was a realistic and workable proposition, which at that time was still unproven.

Just nineteen days after that meeting, the Japanese attacked Pearl Harbor, an event that prompted the WBC committee members to take the whole subject a lot more seriously. The group members performed a search of the available scientific literature relevant to the potential effectiveness of biological weapons. When they issued their report two months later, it was clear that they were convinced by what they had read that the technology, once developed, could be workable. They therefore recommended a full-bore development program, saying in their final conclusion that, "unless the United States is going to ignore this potential weapon, steps should be taken immediately to begin work on the problems of biological warfare."

One of the methods that the original War Bureau of Consultants proposed for inflicting disease upon an enemy was by means of mosquitoes. And why not? Mosquitoes, after all, are known to be very efficient killers. Without harboring any malign intentions toward human beings, female anopheles mosquitoes nevertheless manage to kill about one million people each year by injecting them with the malaria parasite, *Plasmodium falciparum*, in the process of biting. Mosquitoes, nature's own proprietary vehicle of biological warfare, are in effect weapons of mass destruction.

But the War Bureau of Consultants wanted to improve upon nature, and so they recommended that "studies be made to determine whether mosquitoes can be infected with several diseases simultaneously with a view to using these insects as offensive weapons." In its way, this idea was quite ingenious.

Of course, converting this conceptually simple and even fairly obvious idea into a workable, operational offensive weapons system would take some doing, for there were numerous problems to surmount. There was the problem of creating adequate supplies of the insect, enough to infect a substantial portion of a given population. This would require countless millions, or even billions, of insects.

Second was the problem of insuring that all the mosquitoes, or other insect of choice, such as fleas, were harboring the desired disease-causing agent: the yellow fever virus, for example, or the bubonic plague bacterium, or both. And it would be necessary to transport the insects to the target area while they were still alive and well, and then somehow to disperse them on demand without harming the creatures in the process. But such matters were the province of entomologists, and over the course of time the army's entomological warriors would find solutions to these and other difficulties as well.

In 1943, the US Army's Chemical Warfare Service decided on a site at which to conduct their bioweapons research and development project: Detrick Field, which was at that time an abandoned National Guard flying strip located outside of Frederick, Maryland. On March 9, 1943, the US Army Chemical Warfare Service took formal possession of the site, annexed some nearby farmland for field trials, and gave it all the name Camp Detrick (later changed to Fort Detrick).

In the early 1950s, Camp Detrick established an entomology division to work on the problem of converting the mosquito bioweapons delivery scheme from an idea into a practical reality. While the entomological research proceeded at Camp Detrick, the army established an insect production facility elsewhere, at Pine Bluff Arsenal, in Arkansas, which "would house the largest insect-rearing rearing facility in the world, a mosquito mill with the capacity to produce 100 million infected vectors per week."

As with any other weapon, these tiny entomological "bug bombs" would have to be tested first. And so they were, in standard military fashion, in "operations" that had code names that were in some cases all too descriptive: Operation Big Buzz (mosquitoes), Operation Big Itch (fleas), and others.

Big Itch was conducted in 1954, at Dugway Proving Ground. The fleas, which were uninfected, were carried inside cardboard containers that were dropped from aircraft. The containers were attached to parachutes and descended slowly. At a given height, the fleas were released into the open air over a circular array of guinea pigs that were used as test subjects.

Big Buzz, which occurred in May 1955, was not carried out at Dugway but rather in more realistic, operational conditions, which is to say, somewhere in Georgia. For this event, the army bred about a million *Aedes aegypti* mosquitoes (perhaps even Camp Detrick's own strain, the so-called CD *Aedes aegypti*) and stored them for two weeks to simulate real-world travel and delays.

The insects, which were again uninfected, were released from the air over both guinea pigs and human volunteers. Enough of the mosquitoes managed to find these animal and human hosts, despite their being dropped nearly a half mile away from their live targets.

In May 1965, ten years after Operation Big Buzz, the army conducted another such test, at Baker Island in the Pacific. Although Baker was not a typical tropical island, its climate was perhaps similar enough to Vietnam's so that the mosquito trials at Baker could be considered a test case or proof of concept for a possible entomological attack on Vietnam. But so far as is known, no such attacks were ever carried out.

Acknowledgments

For their help in providing me with documents or photographs associated with this story I would first of all like to thank the archivists of the Smithsonian Institution Archives: Deborah Shapiro, Heidi Stover, and Jennifer Morris (Anacostia Community Museum, Smithsonian Institution). Thanks as well to Antony Adler (Carleton College), Steve Berendzen (Arctic National Wildlife Refuge), Autumn-Lynn Harrison (Smithsonian Conservation Biology Institute, National Zoological Park), Krystal Kakimoto (Bishop Museum Archives, Honolulu), Vicki Killian (researcher), and Kate Toniolo (Pacific Remote Islands Marine National Monument).

A few surviving Pacific Project participants were kind enough to answer questions by email, and for their assistance I would like to thank Fred C. Sibley, Max Thompson, James Lewis, and Charles Ely.

Jeff Cox knew Binion Amerson during the early parts of their respective careers, and then again during the later stages of Amerson's life. He provided me with substantial amounts of information about him and sent me more than a hundred photographs taken by Amerson while he was a POBSP field team member. This was a great resource in giving me a vivid impression of what it was like to be out on the islands, and amid their birdlife, on a daily basis. I am extremely grateful to Jeff Cox for all his help.

I also owe a lot to Mark Rauzon (Geography Department, Laney College), who provided me with manuscripts of his unpublished papers on the Pacific Project, gave me good advice and guidance at several points during the composition of this book, and read and provided valuable comments and corrections on four of its core chapters. For all of this and more, many thanks.

Lastly, I am indebted to my editor at Oxford University Press, Jeremy Lewis, for his guidance throughout the project, and for his comments on a late draft of the full manuscript.

The term "invisible history," used in the text, was coined by Richard Preston, in his novel *The Cobra Event*, the second of his books about "dark biology."

Bibliography

References to "RU 245" are to the Smithsonian Institution Archives, Record Unit 245, National Museum of Natural History, Pacific Ocean Biological Survey Program.

Preface

Amerson, Binion. Sand–Johnston Journal. July–August 1963. RU 245 Box 24 Folder 8.

Chapter 1: Secrecy Comes to the Smithsonian

Burleigh, Nina. *The Stranger and the Statesman*. New York: William Morrow, 2003.

Ewing, Heather. *The Lost World of James Smithson: Science, Revolution, and the Birth of the Smithsonian*. New York: Bloomsbury, 2007.

Gup, Ted. "The Smithsonian's Secret Contract: The Link between Birds and Biological Warfare." *Washington Post Magazine*, May 12, 1985, 8–20.

Humphrey, Philip. Email to Ed Regis, December 19, 1998.

Humphrey, Philip. "Proposal of Research to be Undertaken by the Division of Birds, Smithsonian Institution." October 2, 1962. RU 245 Box 8, Folder 20.

Humphrey, Philip. Telephone interview with Ed Regis, December 14, 1998.

National Academy of Sciences. 1980. *Biographical Memoirs*. Vol. 51. Washington, DC: National Academies Press. https://doi.org/10.17226/574.

Regis, Ed. *The Biology of Doom: The History of America's Secret Germ Warfare Project*. New York: Holt, 1999.

Tucker, Jonathan B. *War of Nerves: Chemical Warfare from World War I to Al-Qaeda*. New York: Pantheon, 2006.

US Army Chemical Corps Historical Office. *Summary of Major Events and Problems: United States Army Chemical Corps*. Fiscal Years 1961–1962. Army Chemical Center, Maryland, June 1962.

US Department of the Army. *U.S. Army Activity in the U.S. Biological Warfare Programs*. Vol. 1. February 24, 1977. Unclassified. § 5-5, 5-6.

Whitmore, Frank C., Jr. *Remington Kellogg 1892–1969*. Washington, DC: National Academy of Sciences, 1975.

Web:

U.S. Army Dugway Proving Ground History
https://www.atec.army.mil/dpg/history.html

Chapter 2: Recruitment

Amerson, A. Binion, Jr. Interview with Ed Regis, Dallas, Texas, June 13, 1999.

_____. "The Natural History of French Frigate Shoals, Northwestern Hawaiian Islands." *Atoll Research Bulletin*, no. 150 (December 20, 1971): 1–400.

Clapp, Roger B. "Autobiographical Notes or Fun with Field Work." *Atoll Research Bulletin* 494 (2001): 53–78.

_____. Interview with Ed Regis, Smithsonian Institution, August 31,1998.

Eigelsbach, H. T., and C. M. Downs. "Prophylactic Effectiveness of Live and Killed Tularemia Vaccines: I. Production of Vaccine and Evaluation in the White Mouse and Guinea Gig." *Journal of Immunology* 87 (1961): 415–425.

Ely, Charles. Email to Ed Regis, March 24, 2021.

Ely, Charles. Letter to Research Curators, April 20, 1964. RU 245 Box 12, Folder 4.

Johnston, Richard F. "Ornithology at the University of Kansas." In *Contributions to the History of North American Ornithology*, edited by W. E. Davis Jr. and J. A. Jackson, Memoirs of the Nuttall Ornithological Club, no. 12. Cambridge, Massachusetts (1995): 95–112.

Sibley, Fred C. Email to Ed Regis, February 27, March 24, 2021.

_____. "Pacific Ocean Biological Survey Program." February 2020. Unpublished manuscript.

Thompson, Max. Email to Ed Regis, September 27, 2020, July 2, 2021.

US Department of Agriculture. "United States Veterinary Permit No. 1087 Organisms or Vectors." October 3, 1962. RU 245 Box 4, Folder 10.

US Department of the Interior, Bureau of Sport Fisheries and Wildlife. "Required Procedures to Be Taken Prior to Landing on Wildlife Refuge Islands." December 5, 1962. RU 245 Box 4, Folder 10.

US Department of Land and Natural Resources, State of Hawaii. "Scientific Collecting Permit." January 14, 1963. RU 245 Box 4, Folder 10.

US Fish and Wildlife Service. "Permit to Import Not to Exceed Six Hundred and Ninety-Three (693) Migratory Birds from Islands of the West Pacific." October 11, 1962. RU 245 Box 4, Folder 10.

Web:

Lawrence N. Huber

https://siarchives.si.edu/collections/auth_per_fbr_eacp66

Chapter 3: Prequels

Abulafia, David. *The Boundless Sea: A Human History of the Oceans*. New York: Oxford University Press, 2019.

Beaglehole, J. C. *The Exploration of the Pacific*. 3rd ed. Stanford, CA: Stanford University Press, 1966.

Ely, Charles A., and Roger B. Clapp. "The Natural History of Laysan Island, Northwestern Hawaiian Islands." *Atoll Research Bulletin*, no. 171 (December 31, 1973): 1–361.

Magier, S., and L. Morgan. *A Brief History of Human Activities in the US Pacific Remote Islands*. Seattle, WA: Marine Conservation Institute, 2012.

Olson, Storrs L. "History and Ornithological Journals of the Tanager Expedition of 1923 to the Northwest Hawaiian Islands, Johnston and Wake Islands." *Atoll Research Bulletin*, no. 433 (February 1996): 1–216.

Philbrick, Nathaniel. *Sea of Glory: America's Voyage of Discovery: The U.S. Exploring Expedition, 1838–1842*. New York: Penguin Books, 2003.

Ripley, S. Dillon, and James A. Steed. *Alexander Wetmore 1886–1978*. Washington, DC: National Academy of Sciences, 1987.

Smithsonian Institution Archives. Record Unit 245, National Museum of Natural History, Pacific Ocean Biological Survey Program. Sections: Historical Note, p. 1. Series 31: Southern Grid Survey Reports, p. 63. Series 32: Northern Grid Survey Reports, p. 65. Series 34: Eastern Grid Survey Reports, p. 69, 1979.

Wetmore, Alexander. "Bird Life among Lava Rocks and Coral Sand." *National Geographic Magazine* 158 (July 1925): 77–108.

Web:

United States Exploring Expedition. 2004.
https://www.sil.si.edu/DigitalCollections/usexex/learn/Walsh-01.htm

Chapter 4: Life in the Field

Amerson, Binion. "Leeward & Southern Islands Journal." SIC#1: January 27–March 23, 1963. RU 245 Box 155, Folders 1–4.

Anon. [Roger Clapp?]. "How Smithsonian Bands Pacific Seabirds." *Pacific Bird Observer*, no. 2 (November 1965): 1–2.

Clapp, Roger B. Interview with Ed Regis, Smithsonian Institution, August 31, 1998.
_____. "Journal + Catalogue." SIC#3: 1963. RU 245 Box 155, Folders 7–10.

Ely, Charles A., and Roger B. Clapp. "The Natural History of Laysan Island, Northwestern Hawaiian Islands." *Atoll Research Bulletin*, no. 171 (December 31, 1973): 1–361.

Gerone, Peter J. "Collection of Bird Sera." Fort Detrick, MD: Virology and Rickettsia Division, August 27, 1962.

Huber, Larry. "Journal + Catalogue." SIC#3: October 1963. RU 245 Box 155, Folders 7–10.

Sibley, Fred C. Email to Ed Regis, August 31, 2020.
_____. "Pacific Ocean Biological Survey Program." February 2020. Unpublished manuscript.

Sibley, Fred C., and Roger B. Clapp. "Biological Survey of Howland Island March 1963–May 1965." Washington, DC: Smithsonian Institution, Division of Birds. September 1, 1965. Unpublished report.

Web:

Midway Atoll
https://en.wikipedia.org/wiki/Midway_Atoll
USS *Moctobi* (ATF-105)
http://www.navsource.org/archives/09/39/39105.htm
USS *Tawakoni* (ATF-114)
http://www.navsource.org/archives/09/39/39114.htm

Chapter 5: The Artificial Atoll

Amerson, A. Binion, Jr. *The Coral Carrier: French Frigate Shoals, Northwestern Hawaiian Islands, A History*. Dallas, TX: Binion Amerson Books, 2012.

Amerson, A. Binion, Jr. "The Natural History of French Frigate Shoals, Northwestern Hawaiian Islands." *Atoll Research Bulletin*, no. 150 (December 20, 1971): 1–400.

—————. Sand–Johnston Journal. July–Aug 1963. RU 245 Box 24 Folder 8.

Amerson, A. Binion, Jr., Philip C. Shelton, Roger B. Clapp, and William O. Wirtz II. "The Natural History of Johnston Atoll, Central Pacific Ocean." *Atoll Research Bulletin*, no. 192 (December 1976): 1–479.

Clapp, Roger B. Interview with Ed Regis, Smithsonian Institution, August 31,1998.

Division of Birds, Smithsonian Institution. "Proposal for Amendment of Contract DA-18-064-AMC-56 Entitled Distribution, Ecology, and Migrations of Pacific Birds and Mammals." May 27, 1963. RU 245 Box 8, Folder 21.

Hoerlin, Herman, Los Alamos Scientific Laboratory, and US Energy Research and Development Administration. *United States High-Altitude Test Experiences: A Review Emphasizing the Impact on the Environment*. [Washington, DC]: Energy Research and Development Administration, 1976.

Humphrey, Philip S. "Memorandum: Increase of Scope of SI Program of Research on Birds and Mammals of the Pacific Ocean." May 15, 1963. RU 245 Box 8, Folder 21.

Olson, Storrs L. "History and Ornithological Journals of the Tanager Expedition of 1923 to the Northwest Hawaiian Islands, Johnston and Wake Islands." *Atoll Research Bulletin*, no. 433 (February 1996): 1–216.

Pacific Bird Observer, no. 1 (September 1965).

Quaile, John E. "French Frigate Shoals." *Military Engineer* 39 (1947): 383.

Rauzon, Mark J. *Isles of Amnesia: The History, Geography, and Restoration of America's Forgotten Pacific Islands*. Honolulu: University of Hawai'i Press, 2016.

Web:

NOAA. "Assessing the Damage: The First Step After Hurricane Walaka." 2018.
https://www.papahanaumokuakea.gov/new-news/2018/11/21/ffs-walaka/

Executive Order No. 4467. 1926.
https://en.wikisource.org/wiki/Executive_Order_4467.

Google Street View, French Frigate Shoals. 2013.
https://tinyurl.com/uxshbycc

Shady Grove Weapons Test. 2003.
https://www.health.mil/Reference-Center/Fact-Sheets/2003/12/02/Shady-Grove-Revised

Chapter 6: Project 112

Buhlman, Ernest H. Test 64-4—SHADY GROVE. Final Report. Report No. DTC 644115R. Department of the Army. Fort Douglas, Utah: Deseret Test Center, June 1966. (A selection of redacted pages from this partially declassified, originally secret document may be found at the "Black Vault" Web site: https://static.secure.website/wsc fus/10426050/7109108/1018.pdf, at 383–412.)

Hewes, James E., Jr. *From Root to McNamara: Army Organization and Administration.* Washington, D.C.: Center of Military History, United States Army, 1975.

Huber, L. N. Journal. SIC#7: Jan 26–Mar 27, 1965. RU 245 Box 26, Folder 7. https://www.biodiversitylibrary.org/bibliography/138987.

Tsiodras, S., Theodoros Kelesidis, Iosif Kelesidis, Ulf Bauchinger, and Matthew E. Falagas. "Human Infections Associated with Wild Birds." *Journal of Infection* 56 (2008): 83–98. doi:101016/j.jinf.2007.11.001.

US Army Chemical Corps Historical Office. *Summary of Major Events and Problems: United States Army Chemical Corps: Fiscal Years 1961–1962.* Army Chemical Center, Maryland, June 1962.

US Department of the Army. *U.S. Army Activity in the U.S. Biological Warfare Programs.* Vol. 2. February 24, 1977. Unclassified. Annex I.

Web:

National Security Action Memorandum No. 235. 1963. https://www.jfklibrary.org/asset-viewer/archives/JFKNSF/340/JFKNSF-340-023

Chapter 7: "Bird Bombs"

Brown, C. A., and V. J. Cabelli. *Summary Report on the Susceptibility of Birds to Tularemia: The Wedge-Tailed Shearwater and the Black-Footed Albatross.* Dugway Proving Ground, UT: October 1964. Unclassified.

Clapp, Roger B. "The Natural History of Gardner Pinnacles, Northwestern Hawaiian Islands." *Atoll Research Bulletin,* no. 163 (1972): 1–25.

Miller, William S., Charles R. Rosenberger, Robert L.Walker, and Edwin C. Corristan. "Susceptibility of Sooty Terns to Venezuelan Equine Encephalitis (VEE) Virus." Technical Manuscript 99. Fort Detrick, MD: US Army Biological Laboratories, October 1963. Unclassified.

Stone, Roger D. *The Lives of Dillon Ripley: Natural Scientist, Wartime Spy, and Pioneering Leader of the Smithsonian Institution.* Lebanon, NH: ForeEdge, 2017.

Chapter 8: The Military Payoff

Buhlman, Ernest H. *Test 64-4—SHADY GROVE: Final Report.* Report No. DTC 644115R. Department of the Army. Fort Douglas, UT: Deseret Test Center, June 1966. (A selection of redacted pages from this partially declassified, originally secret document may be found at the "Black Vault" website: https://static.secure.website/wscfus/10426050/7109108/1018.pdf. [See 383–412.])

Deseret Test Center (DTC). *Test 65-4—MAGIC SWORD.* Fort Douglas, UT: Deseret Test Center, May 1966.

Husted, Dayle. Trip Report. Miscellaneous Pelagic Cruise No. 5. Honolulu to Guam. June 2 to 27, 1966. USCGC *Basswood.* RU 245 Box 17, Folder 10. https://www.biodiversitylibrary.org/bibliography/142301.

Lockwood, Jeffrey A. *Six-Legged Soldiers: Using Insects as Weapons of War.* New York: Oxford University Press, 2009.

Morrison, John H. *DTC Test 68-50: Test Report.* Vol. 1. Fort Douglas, UT: Deseret Test Center, March 1969.

Sidell, Frederick R., Ernest T. Takafuji, and David R. Franz. *Medical Aspects of Chemical and Biological Warfare.* Textbook of Military Medicine. Part 1: Warfare, Weaponry, and the Casualty. Washington, DC: Walter Reed Army Medical Center, 1997.

Spendlove, J. Clifton. "The Time of My Life: A Personal History." Salt Lake City, UT: LDS Archives, 1994. Unpublished memoir.

The Department of Defense's Inquiry into Project 112/Shipboard Hazard and Defense (SHAD) Tests before the Senate Subcommittee on Personnel of the Committee on Armed Services. 107th Cong. 2nd sess (October 10, 2002).

Web:

USS *Granville S. Hall* (YAG-40)
https://en.wikipedia.org/wiki/USS_Granville_S._Hall_(YAG-40)

Chapter 9: The "Secret" Emerges

Boffey, Philip M. "Biological Warfare: Is the Smithsonian Really a 'Cover'?" *Science* 163 (February 21, 1969): 791–796.

Gould, Stephen Jay. "Smithsonian's Albatross." *Science* 164 (May 2, 1969): 497.

Gup, Ted. "The Smithsonian's Secret Contract: The Link between Birds and Biological Warfare." *Washington Post Magazine* (May 12, 1985): 8–20.

Henson, Pamela M. "The Smithsonian Goes to War: The Increase and Diffusion of Scientific Knowledge in the Pacific." In *Science and the Pacific War,* edited by Roy M. McLeod, 27–50. Kluwer Academic, 2000.

Humphrey, Philip S. "An Ecological Survey of the Central Pacific." In *Smithsonian Year 1965.* Washington, DC: Smithsonian Institution, 1965: 24–30.

MacLeod, Roy. "'Strictly for the Birds': Science, the Military and the Smithsonian's Pacific Ocean Biological Survey Program, 1963–1970." *Journal of the History of Biology* 34 (2001): 315–352.

Sibley, Fred C. Email to Ed Regis, March 24, 2921.

Siekevitz, Philip. Letter to *Scientific Research* (January 6, 1969): 6–7.

Small, William E. "DOD Supporting Bird Studies in Pacific, Brazil." *Scientific Research,* December 9, 1968: 27.

"Smithsonian Bird Research Tied to Germ Warfare Study." *Washington Post,* February 5, 1969.

Thompson, Max C., and Robert L. DeLong. "The Use of Cannon and Rocket-Projected Nets for Trapping Shorebirds." *Bird-Banding* 38 (1967): 214–218.

Tucker, Jonathan B., and Erin R. Mahan. *President Nixon's Decision to Renounce the U.S. Offensive Biological Weapons Program.* Washington, DC: National Defense University Press, 2009.

Web:

Nixon Statement. 1969.
https://2001-2009.state.gov/r/pa/ho/frus/nixon/e2/83597.htm
Kissinger Memorandum. 1969.
https://2001-2009.state.gov/r/pa/ho/frus/nixon/e2/83569.htm

Chapter 10: Fate of the Islands

Berendzen, Steve. Email to Ed Regis, July 13, 2020.

Berendzen, Stephen L., and Douglas J. Forsell. "Howland Island National Wildlife Refuge Expedition Report." March 25–April 12, 1986. Unpublished report. USFWS, Honolulu, HI.

King, Warren B. "Conservation Status of Birds of Central Pacific Islands." *Wilson Bulletin* 85 (March 1973): 89–103.

The Radiological Cleanup of Enewetak Atoll. United States: Defense Nuclear Agency, 1981.

Rauzon, Mark J. *Isles of Amnesia: The History, Geography, and Restoration of America's Forgotten Pacific Islands.* Honolulu: University of Hawai'i Press, 2016.

Report on the Status of the Runit Dome in the Marshall Islands. Report to Congress. Washington, DC, June 2020.

Rust, Suzanne. "How the U.S. Betrayed the Marshall Islands, Kindling the Next Nuclear Disaster." *Los Angeles Times,* November 10, 2019.

Schreiber, E. A. *Breeding Biology and Ecology of the Seabirds of Johnston Atoll, Central Pacific Ocean: Results of a Long-Term Monitoring Project 1984-2003.* Aberdeen, MD: Aberdeen Proving Ground, 2018.

Sibley, Fred C., and Roger B. Clapp. "Biological Survey of Howland Island March 1963– May 1965." Washington, DC: Smithsonian Institution, Division of Birds. September 1, 1965. Unpublished report.

Web:

Johnston Island Real Estate. 2006.
http://great-hikes.com/blog/island-for-sale/
Unusual Real Estate Listing # 6384
Johnston Island
http://www.clui.org/sites/default/files/29_winter_2006.pdf (20–21)

Chapter 11: Aftermath and Aftereffects

Amerson, A. Binion, Jr. Interview with Ed Regis, Dallas, Texas, June 13, 1999.

Church, Jane P. Telephone interview with Ed Regis, July 25, 1999.

Clapp, Roger B. "Autobiographical Notes or Fun with Field Work." *Atoll Research Bulletin,* Vol. 494 (2001): 53–78.

_____. "Notes on the Birds of Kwajalein Atoll, Marshall Islands." *Atoll Research Bulletin,* no. 342 (1990): 1–94.

Collins, Michael. *Carrying the Fire: An Astronaut's Journeys.* New York, Farrar, Straus and Giroux, 1974.

Committee on Shipboard Hazard and Defense II (SHAD II). *Assessing Health Outcomes Among Veterans of Project SHAD.* Institute of Medicine. Washington, DC: National Academies Press, 2016.

Crossin, Richard. Telephone interview with Ed Regis, July 25, 1999.

Deseret Test Center. 1965. *Test 64-2—Flower Drum (U), Phase I: Final Report—Revised.* December.

Gup, Ted. "The Smithsonian's Secret Contract: The Link between Birds and Biological Warfare." *Washington Post Magazine*, May 12, 1985, 8–20.

Harrison, Autumn-Lynn. "An Ode to the Pacific Ocean Biological Survey Program and Its Biologists, 1963–1969." *Seabirds: Global Ocean Sentinels*. Second World Seabird Conference (Cape Town, South Africa, October 26–30, 2015).

Holloway, Karel. "Daylily Fancier Who Lost Home and Garden Gives Collection to Dallas Suburb." *Dallas Morning News*, June 18, 2014.

Huber, Lawrence N. "Notes on the Migration of the Wilson's Storm Petrel *Oceanites oceanicus* Near Eniwetok Atoll Western Pacific Ocean." *Notornis* 18 (March 1971): 38–42.

Kakesako, Gregg K. "Ex-HPD Officer Victim of Military Chemical Tests." *Star-Bulletin*, October 11, 2002.

Page, William F., et al. *Long-Term Health Effects of Participation in Project SHAD*. Institute of Medicine. Washington, DC: National Academies Press, 2007.

Schweitzer, Ally. "D.C.'s Least-Visited Smithsonian, the Anacostia Community Museum, Is Preparing for Gentrification." wamu.org. Local News. (September 26, 2017). https://wamu.org/story/17/09/26/d-c-s-least-visited-smithsonian-anacostia-community-museum-preparing-gentrification/

Shanker, Thom, and William J. Broad. "Sailors Sprayed with Nerve Gas in Cold War Test, Pentagon Says." *New York Times*, May 24, 2002, 1.

Sibley, Fred C. Email to Ed Regis, February 27, 2021.

Smithsonian Institution. "Smithsonian to Celebrate Groundbreaking 'FUTURES' Exhibition for Its 175th Anniversary." News release, February 23, 2021.

The Department of Defense's Inquiry into Project 112/Shipboard Hazard and Defense (SHAD) Tests Before the Senate Subcommittee on Personnel of the Committee on Armed Services. 107th Cong. 2nd sess. (October 10, 2002).

Web:

Team Infidel. International Military Forums, June 13, 2008. (Transcript of CBS Evening News with Katie Couric, June 12, 2008).
https://www.military-quotes.com/forum/retired-navy-officer-seeks-justice-t63778.html
Project SHAD Fact Sheets
https://www.health.mil/Military-Health-Topics/Health-Readiness/Environmental-Exposures/Project-112-SHAD/Fact-Sheets
Institute of Medicine: Brief Descriptions of SHAD Tests. 2016.
https://www.nap.edu/read/21846/chapter/5

Epilogue

Ewing, Heather. *The Lost World of James Smithson: Science, Revolution, and the Birth of the Smithsonian*. New York: Bloomsbury, 2007.
Unusual Real Estate Listing # 6384
Johnston Island
http://www.clui.org/sites/default/files/29_winter_2006.pdf (20–21)

Appendix I

US Army Biological Laboratories. "Studies on the Distribution, Ecology, and Immigrations of Pacific Birds and Mammals." POBSP Contract. Frederick, MD: Fort Detrick, October 26, 1962. RU 245 Box 8, Folder 20.

Appendix II

Lockwood, Jeffrey A. *Six-Legged Soldiers: Using Insects as Weapons of War*. New York: Oxford University Press, 2009.

Regis, Ed. *The Biology of Doom: The History of America's Secret Germ Warfare Project*. New York: Holt, 1999.

Index